Energy, Society and Environment

Society's use of energy is at the heart of many environmental problems we face at the beginning of the twenty-first century, most notably global warming and climate change. *Energy, Society and Environment* takes a critical look at both current energy use and possible alternatives, such as the expansion of the use of wind and other renewable energy sources, and nuclear power.

Energy, Society and Environment explores the ways in which energy interacts with society and the environment. The book is structured to provide:

- an understanding of energy-related environmental problems
- an appreciation of the strengths and weaknesses of technological solutions
- knowledge of the social and institutional obstacles to implementing these
- an understanding of the strategic issues facing sustainable energy use

The revised edition reflects recent changes in the area. Chapters on nuclear and wind energy have been revised in response to recent debates. Coverage of fossil fuels has also been strengthened, whilst there is greater emphasis on environmental and energy policy in the context of the debate surrounding the Kyoto accord. Additional case studies have been added which highlight alternative energy solutions.

Energy, Society and Environment examines the potential and limits of technological solutions to energy-related environmental problems and suggests that social, economic and political solutions may also be necessary to avoid serious environmental damage in the future. Global case studies are used throughout to ground the debates and illustrate the interaction between technological and social aspects.

David Elliott is Professor of Technology Policy in the Open University Faculty of Technology and Director of the OU Energy and Environment Research Unit.

Routledge Introductions to Environment Series
Published Titles

Titles under Series Editors:
Rita Gardner and A.M. Mannion

Titles under Series Editor:
David Pepper

Environmental Science texts

Atmospheric Processes and Systems
Natural Environmental Change
Biodiversity and Conservation
Ecosystems
Environmental Biology
Using Statistics to Understand the
    Environment
Coastal Systems
Environmental Physics
Environmental Chemistry

Environment and Society texts

Environment and Philosophy
Environment and Social Theory
Energy, Society and Environment,
    2nd edition
Environment and Tourism
Gender and Environment
Environment and Business
Environment and Politics, 2nd edition
Environment and Law
Environment and Society

Routledge Introductions to Environment Series

Energy, Society and Environment

Technology for a sustainable future
Second edition

David Elliott

Routledge
Taylor & Francis Group

LONDON AND NEW YORK

First published 1997
by Routledge
11 New Fetter Lane, London EC4P 4EE

Simultaneously published in the USA and Canada
by Routledge
29 West 35th Street, New York, NY 10001

Reprinted 1998

Second edition 2003

Routledge is an imprint of the Taylor & Francis Group

Typeset in Times by Keystroke, Jacaranda Lodge, Wolverhampton
Printed and bound in Great Britain by The Cromwell Press,
Trowbridge, Wiltshire

British Library Cataloguing in Publication Data
A catalogue record for this book is available from the British Library

Library of Congress Cataloging in Publication Data
Elliott, David, 1943–
    Energy, society & environment / David Elliott. – 2nd ed.
        p. cm. – (Routledge introductions to environment series)
    Includes bibliographical references and index.
    1. Power resources–Environmental aspects. 2. Technology–Social aspects.
    I. Title: Energy, society, and environment. II. Title. III. Series.
    TD195.E49E42 2003
    333.79′14–dc21   2003000824

ISBN 0–415–30485–7 (hbk)
ISBN 0–415–30486–5 (pbk)

For Oliver, Olga and Tam and in memory of Ruth and Don

Contents

# Figures

 **Tables**

Boxes

Series editor's preface
Environment and Society titles

The modern environmentalist movement grew hugely in the last third of the twentieth century. It reflected popular and academic concerns about the local and global degradation of the physical environment which was increasingly being documented by scientists (and which is the subject of the companion series to this, *Environmental Science*). However it soon became clear that reversing such degradation was not merely a technical and managerial matter: merely knowing about environmental problems did not of itself guarantee that governments, businesses or individuals would do anything about them. It is now acknowledged that a critical understanding of socio-economic, political and cultural processes and structures is central in understanding environmental problems and establishing environmentally sustainable development. Hence the maturing of environmentalism has been marked by prolific scholarship in the social sciences and humanities, exploring the complexity of society–environment relationships.

Such scholarship has been reflected in a proliferation of associated courses at undergraduate level. Many are taught within the 'modular' or equivalent organizational frameworks which have been widely adopted in higher education. These frameworks offer the advantages of flexible undergraduate programmes, but they also mean that knowledge may become segmented, and student learning pathways may arrange knowledge segments in a variety of sequences – often reflecting the individual requirements and backgrounds of each student rather than more traditional discipline-bound ways of arranging learning.

The volumes in this *Environment and Society* series of textbooks mirror this higher educational context, increasingly encountered in the early twenty-first century. They provide short, topic-centred texts on social science and humanities subjects relevant to contemporary society–environment relations. Their content and approach reflect the fact that each will be read by students from various disciplinary backgrounds, taking in not only social sciences and humanities but others such as physical and natural sciences. Such a readership is not always familiar with the disciplinary background to a topic, neither are readers necessarily going on to further develop their interest in the topic. Additionally, they cannot all automatically be thought of as having reached a similar stage in their studies – they may be first-, second- or third-year students.

The authors and editors of this series are mainly established teachers in higher education. Finding that more traditional integrated environmental studies and specialized texts do not always meet their own students' requirements, they have often had to write course materials more appropriate to the needs of the flexible undergraduate programme. Many of the volumes in this series represent in modified form the fruits of such labours, which all students can now share.

Much of the integrity and distinctiveness of the *Environment and Society* titles derives from their characteristic approach. To achieve the right mix of flexibility, breadth and depth, each volume is designed to create maximum accessibility to readers from a variety of backgrounds and attainment. Each leads into its topic by giving some necessary basic grounding, and leaves it usually by pointing towards areas for further potential development and study. There is introduction to the real-world context of the text's main topic, and to the basic concepts and questions in social sciences/humanities which are most relevant. At the core of the text is some exploration of the main issues. Although limitations are imposed here by the need to retain a book length and format affordable to students, some care is taken to indicate how the themes and issues presented may become more complicated, and to refer to the cognate issues and concepts that would need to be explored to gain deeper understanding. Annotated reading lists, case studies, overview diagrams, summary charts and self-check questions and exercises are among the pedagogic devices which we try to encourage our authors to use, to maximize the 'student friendliness' of these books.

Hence we hope that these concise volumes provide sufficient depth to maintain the interest of students with relevant backgrounds. At the same time, we try to ensure that they sketch out basic concepts and map their territory in a stimulating and approachable way for students to whom the whole area is new. Hopefully, the list of *Environment and Society* titles will provide modular and other students with an unparalleled range of perspectives on society–environment problems: one which should also be useful to students at both postgraduate and pre-higher education levels.

David Pepper

May 2000

Series International Advisory Board

Australasia: Dr P. Curson and Dr P. Mitchell, Macquarie University

North America: Professor L. Lewis, Clark University; Professor L. Rubinoff, Trent University

Europe: Professor P. Glasbergen, University of Utrecht; Professor van Dam-Mieras, Open University, The Netherlands

 # **Preface**

Energy use is fundamental to human existence and it should come as no surprise that the way mankind has been using it is at the heart of many of the environmental problems that have emerged in recent years. There are many types of pollution, but the emissions from the combustion of fuels in power stations and cars are probably the most worrying for many people, given the impact of air quality on health. More generally, the use of fossil fuels such as coal, oil and gas is increasingly seen as having major global environmental impacts, such as global warming. There are also major concerns over the risk of release of radioactive materials associated with the use of nuclear fuels. These energy sources also underpin what many see as an unsustainable form of industrial society, in which the environment is treated as a 'free' resource of energy and other materials and as a more or less infinite sink for wastes, with the scale and pace of the 'throughput' from source to sink growing ever greater.

This book explores the way in which energy use interacts with society and the environment, but the emphasis is not so much on describing the problems as on looking at some possible strategic solutions. Some of the solutions involve new technology: our use of technology has been a major cause of environmental problems, but it is sometimes argued that technology can be improved and used more wisely so as to avoid them in future. However, there are also limits to what technology can do: in the end it may be that there will also be a need for social, economic and political change if serious environmental damage is to be avoided. Part of the aim of this book is to try to explore just how far purely technical solutions can take us, and then to look at the strategic alternatives.

This is not a source book of environmental problems or technical solutions. The emphasis is mainly on social processes and strategic issues, rather than on technical details. I have however provided some technological background where it is needed. There are also Further Reading guides at the end of each chapter, pointing to key texts, some of which look at the technological aspects in more detail. Equally, this is not a social science textbook. I have located the discussion of social issues within a technological context, through case studies and examples, to try to make the issues more concrete, and to illustrate the interaction between the technological and social aspects. The boxes scattered through the text amplify these points via specific examples and case studies. I have also included some questions in Appendix I to help you consolidate and develop your understanding of the arguments.

Inevitably, given the space limitations, there have had to be some omissions. For example, I have not been able to cover the transport issue in any depth. That deserves a book of its own. In addition, the emphasis in the case studies is as mainly on developments in the industrialised countries, although I have attempted to cover general trends worldwide and to set the discussion in an international context.

The structure of the book

This book is structured in four main parts. In the first part there is a general introduction to the key environmental issues and problems. Then, in Part 2, there is a review of some of the key technological solutions, followed, in Part 3, by a review of some of their implementation problems. Finally, in Part 4 there is a discussion of the wider implications for society of attempting to develop a sustainable approach to energy use.

The first task, in Part 1, is to get an idea of the problems the world faces, by looking at the way energy is used and at the way energy use and the environment interact. Energy use is of course primarily related to other human activities – such as heating homes, moving objects and people, manufacturing things and growing food. In each case technologies have been developed which use fuels to provide power. So the initial survey in Chapter 1 looks quite broadly at the technology used and at the forces shaping its development, and sets out a conceptual model of interactions between technology, society and the environment.

Then Chapter 2 looks at the environmental implications of the use of technology, with energy use as the key issue. This analysis of the environmental problems associated with existing forms of energy generation and use leads on, in Chapter 3, to a basic set of criteria for environmentally sustainable energy technology.

With the survey of problems as a background, and armed with the criteria for sustainable technology, Part 2 looks critically at some possible technological solutions to energy-related environmental problems, some of which are seen as relatively limited 'technical fixes', in that they only deal with symptoms rather than causes. For example, many technical fixes are essentially post hoc 'remedial' measures, attempting to 'clean up' harmful emissions – the so-called 'end-of-pipe' approach. Thus flue-gas scrubbing devices are used to filter out harmful emissions from power stations and catalytic converters are added to the end of car exhaust pipes.

There are obviously limits to this sort of approach: some emissions cannot be easily or economically filtered out, notably the carbon dioxide gas which is a fundamental product of combustion and a major contributor to the greenhouse 'global warming' effect. Some fossil fuels produce less carbon dioxide than others, so there is some potential for reducing emissions by switching fuels, but in the end the most direct way to reduce carbon dioxide production is to burn less fossil fuel. One response is therefore to try to reduce demand for energy by avoiding waste through the adoption of energy conservation measures at the point of use, like insulation in buildings, and by developing more efficient energy-using technology. In the last few years there have also been moves to develop 'greener' products and 'cleaner' production processes, and one of the aims has been to improve the efficiency with which they use energy. Some people believe that a shift to a 'conserver society' is a vital element in any attempt to live in an environmentally sustainable way and Chapter 4 looks at the basic technical elements of this approach.

However, although energy conservation has enormous potential, there will still be a need to generate some energy. Assuming that the use of fossil fuel must be reduced and perhaps eventually eliminated, the main alternative new energy supply options are nuclear power and renewable energy technology – energy derived from natural sources like the wind, waves and tides. Chapters 5 and 6 look at nuclear power. Once seen as the energy source of the future, the technical, economic, environmental and social problems seem to have multiplied, while the attractions of the alternatives to nuclear – such as the use of natural renewable energy sources like wind power – seem to have increased. The conclusion would seem to be that, although there are still some strong supporters, nuclear power is unlikely to meet our criteria, while renewable energy, if coupled with energy conservation, seems likely to be a viable option – at least technically. Chapter 7 therefore looks at the renewables in detail, reviewing the basic technology. Chapter 8 then looks at renewable energy developments around the world and Chapter 9 rounds off our survey of renewables with a discussion of development strategy.

Having set out the potentially sustainable energy supply option of renewable energy, Part 3 attempts to look at the problems relating to the deployment of renewable energy and other sustainable energy technologies. Chapters 10 and 11 look at some of the institutional obstacles and implementation problems, while Chapter 12 presents a case study of public reactions to the deployment of wind farms in the UK. This case study is analysed in Chapter 13.

Part 4 widens the focus and asks, assuming that the obstacles discussed in Part 3 can be overcome, to what extent can the various solutions discussed in Part 2 contribute to genuine environmental sustainability. Chapter 14 asks can we just rely on technical fixes or are social and political changes also needed? Chapter 15 reminds us that a global perspective is needed, and that this raises issues concerning world economic development.

Clearly, our discussion must touch on some very broad and potentially contentious issues concerning for example consumerism, economic redistribution and political power, which no one book can hope to resolve. However, Chapter 16 attempts to round off our analysis by looking at the tactical and strategic social, technological and the environmental choices we face, while Chapter 17 summarises the overall conclusions and looks at some possible ways ahead.

The second edition has given me the opportunity to bring the account of the development of sustainable energy technology up to date. However, although a lot of technological progress has been made, the basic political and social choices remain the same. Indeed, if anything, they are now more urgent.

Study notes

It is assumed that some readers will be using this book as part of a study programme in the context of higher education. These study notes are designed to provide an introductory educational guide.

Aims

The Preface sets out the overall rationale and structure of the book, but students may find it helpful to have a more formal checklist of aims and objectives.

Basically, **Part 1** of this book should help you to develop an overview of energy-related environmental issues, including an appreciation of how they have emerged historically. More specifically Part 1 should help you to:

- appreciate that the interaction between the various parts of human society and the rest of the ecosystem is complex and that human energy needs and demands play a major part in shaping this interaction (Chapters 1 and 2);
- understand the arguments for the adoption of more sustainable approaches to energy generation and use, and for the development of criteria for selecting the appropriate technologies (Chapter 3).

Part 2 is designed to help you to appreciate the strengths and weaknesses of the various sustainable energy options, including energy conservation and the 'alternative' (i.e. non-fossil) energy supply options.

More specifically it should help you to:

- appreciate the potential and limitations of 'technical fixes' as a means of resolving environmental problems, and the need for more radical solutions (Chapter 4);

- analyse the pros and cons of nuclear power (Chapter 5 and 6);
- understand the nature and potential of renewable energy technology (Chapter 7);
- appreciate some of the ways in which renewable energy has been developed and deployed around the world (Chapter 8);
- understand some of the strategic development issues facing sustainable energy technology (Chapter 9).

Part 3 is designed to help you explore some of the institutional and social problems facing the development and use of sustainable energy technology. More specifically it should help you to:

- appreciate the institutional problems facing novel energy technologies seeking initial research support, e.g. from governments (Chapter 10);
- appreciate the institutional problems facing novel energy technologies in obtaining finance for large-scale deployment (Chapter 11);
- understand the problem of winning public acceptance for novel technologies, as exemplified by local reactions to wind farms (Chapter 12);
- appreciate the need to negotiate trade-offs between the 'local costs' and the 'global benefits' of renewable energy technologies, in the context of winning public acceptance for them (Chapter 13).

Part 4 is designed to broaden the discussion to help you understand some of the key strategic issues facing sustainable energy technology. More specifically it should help you to:

- appreciate that there are differing views about what sustainability is, and whether it is needed, or obtainable (Chapter 14);
- understand some of the ways in which technological changes, the industrialisation process, and economic development patterns have interacted historically around the world – and some of the environmental limits which might constrain continued economic growth (Chapter 15);
- appreciate that there are no unique blueprints for sustainability, but that the future has to be negotiated to try to avoid inequitable developments and devise an acceptable pattern of use of technology (Chapter 16);
- appreciate how attempts have been made to develop and negotiate criteria for sustainability, and the various strategies that have emerged for bringing about change (Chapter 17).

Learning aids

I assume that you will read through this text sequentially. But to aid your navigation around the text, and to help you consolidate your understanding, as well as the index, I have provided 'topic summaries' at the beginning of each chapter and 'summary points' at the end of each chapter. There are also some general questions (and answers) at the end of the book, and a glossary of key terms and abbreviations follows later in this section.

Non-technical readers will hopefully find the introduction to basic energy terms and units in Chapter 2 helpful, while the model of social–environmental interactions in Chapter 1 may be useful for non-social science students. I refer back to it regularly in the rest of the book, particularly in Part 4.

Preliminary ques tions

As a preliminary exercise before starting the book, you might like to consider the following questions. Some short notes on them are also included, but write down your reactions to the questions before looking at the notes. You might like to come back to the questions, the notes and your answers after you have finished the book, to see to what extent you have developed your understanding, or changed your views.

Questions

1 To what extent does your own personal lifestyle affect the planet in terms of the energy you use at work and play?
2 To what extent could you reduce any adverse environmental impacts from your use of energy?
3 What do you think will be the key energy sources of the future?
4 Can the whole world adopt Western-style consumption levels (i.e. of goods and services)?
5 Can the rich industrial world cut back on its level of consumption and should it do so?
6 Are you optimistic about the future or do you fear that environmental problems will get worse?
7 What local, national and global actions do you think can help secure a sustainable future?

Notes

1   As individuals, at least in the developed countries, we benefit from the use of electricity, gas and petrol but we are becoming aware that this is causing environmental problems. Domestic heating and car use, along increasingly with air transport, are the major 'personal' uses of energy, but the provision of goods and services also involves very significant energy use, over which individual consumers have little influence.

2   Individual lifestyle changes can help reduce energy use (e.g. less car use), as can investing in more energy-efficient homes and domestic devices. But to have a major impact on energy use there will have to be wider changes – in energy generation technology, industry and possibly also in society.

3   The fossil fuels (coal, oil, gas) will not last forever. Nuclear power and natural renewable energy sources are the main alternatives – along with energy conservation. However, not everyone is happy to consider nuclear power as a safe option, and there may be problems with relying on renewables.

4   Industrialisation of the sort pioneered in the West is unlikely to be environmentally sustainable on a global scale. The developing and developed countries may have to adopt alternative technological approaches – and even then it is not clear if economic growth of the current sort can continue worldwide.

5   It is not clear whether humanity can adapt quickly enough in order to develop sustainable approaches to living on this planet – and some say we do not need to anyway. But should it be thought necessary, the opportunity for change exists, assuming vested interests in the status quo can be overcome.

6   There is no one single way forward. Diversity is a good ecological principle, but if we wish to avoid destructive competition for scarce resources, we will have to learn how to co-operate and negotiate more effectively.

Further study

This is only an introductory book, and many of the ideas and issues have inevitably been simplified. I have provided guides to further reading at the end of each chapter. There are also extensive references in the text.

Facts and values

Finally, a word of warning. While I have tried to back up the arguments and analysis in this book with references, and to present the factual material as objectively as possible, some of the issues discussed are controversial. I have tried to strike a balance between extreme views and complacent views, and between optimistic and pessimistic interpretations, but inevitably biases will creep in. One of the skills you will need to develop in approaching this subject is to read critically and, so far as is possible, distinguish between facts and value judgements.

While I hope that I have presented a reasonably balanced account, not everyone will agree. This book does after all discuss challenges to the status quo. Some people may argue that the analysis is too radical or too partisan. Others may suggest that it is not radical enough. You will have to make up your own mind – and one way to try to do that is by reading other books written from other perspectives. That is why I have tried to ensure that some divergent views are represented in the Further Readings.

Acknowledgements

An early version of parts of Chapter 3 appeared in a paper I co-authored with Alexi Clarke in the *International Journal of Global Energy Issues* (9(4/5), 1996). Chapters 6 and 7 owe much to the chapter I wrote on nuclear power for the OU book produced for T206: Energy for a Sustainable Future, *Energy Systems and Sustainability* (Oxford University Press, 2003). Chapters 8, 10, 11 and 15 draw on material I developed for Blocks 5 and 7 of the OU course on Innovation: Design, Environment and Strategy (T302), 1996. Chapter 12 is based on parts of a paper I produced for the House of Commons Welsh Affairs Select Committee (Second Report, Session 1993–4, vol. III, pp. 432–42, HMSO, London, 1994), a revised version of which was subsequently published as 'Public reactions to windfarms: the dynamics of opinion formation', in *Energy and Environment* (5(4): 343–62, 1994).

Versions of some of the material in this book have also appeared in *RENEW*, the journal I edit for NATTA.

Thanks are due to Tam Dougan for her critical support. She and our son Oliver not only coped with my preoccupation while writing the original version but also had to cope with the updating exercise.

Sally Boyle from the OU Department of Design and Innovation, aided by Richard Hearne, provided the necessary skills to convert the graphics into understandable form, and Melanie Attridge and the team at Routledge converted the text into what you have now.

However, responsibility for the final product rests with the author. In writing this book, designed to be accessible to non-technical people, I have inevitably had to try to simplify some complex technological issues. I trust practitioners will bear with me on this, although I would welcome comments and criticism. Although I have tried to be even-handed and to subject all the technologies and policies discussed to critical assessment, inevitably biases exist. This seems unavoidable if one is attempting to discuss alternatives to the status quo, but I would welcome comments on any part of the analysis that is felt to be misleading or incorrect.

The publishers and I would like to thank the following for granting permission to reproduce material in this work: Philip Allan Updates for Figures 1.1 and 4.2; Worldwatch Institute for Figures 1.2, 1.3, 1.4 and 15.3; James and James for Figure 3.1; David Olivier for Figure 4.1; AEA Technology for Figure 7.2; MCT for Figure 7.4; Shell for Figures 9.1, 9.2 and 9.3; Environment Information Services for Figure 12.1; Greenpeace for Figure 14.1; Elsevier Science for Figure 15.2. Every effort has been made to trace copyright holders, but in a few cases this has not been possible. Any omissions brought to our attention will be remedied in future editions.

Abbreviations

CCGT	combined cycle gas turbines
CHP	combined heat and power (co-generation)
CPRE	Council for the Protection of Rural England
DNC	declared net capacity
DTI	Department of Trade and Industry
EC	European Commission (of the European Union)
EST	Energy Saving Trust
ETSU	Energy Technology Support Unit
FBR	fast breeder reactor
GW	gigawatt (1,000 megawatts)
ICT	information and communication technologies
kWh	kilowatt hour (1,000 watts for one hour)
MW	megawatt (1,000 kilowatts)
NFFO	Non-Fossil Fuel Obligation
PV	photovoltaics (solar cells)
PWR	pressurised water reactor
REC	regional electricity company
ROC	Renewables Obligation Certificate
SRC	short rotation coppicing
TWh	terawatt-hour (1,000,000,000 kWhs)
UKAEA	United Kingdom Atomic Energy Authority

 **Environmental problems**

Part 1 reviews the way in which mankind's use of energy impacts on the environment. It looks at the way in which energy use has grown dramatically since the industrial revolution and at some of the key environmental problems that have emerged. It also introduces the idea of sustainable development as an alternative approach and sets out some basic environmental criteria for sustainable energy technologies.

1 Technology and society

- Energy use in society
- Environmental impacts
- Sustainable development
- Negotiating interactions

This introductory chapter sets the scene by looking at the interaction between people and the planet, with the focus on energy use. The ever-increasing pattern of energy use seems unlikely to be environmentally sustainable, in which case we will need to try to negotiate a new way forward. To try to describe some key features of the human–environmental interaction, and how it might be modified, this chapter introduces an analytical model of the various conflicting interests, which is used throughout the book.

People and the planet

Human beings have developed a capacity to create and use tools – or what is now called technology. Technology provides the means for modifying the natural environment for human purposes – providing basic requirements like shelter, food, warmth, as well as communications, transport and a range of consumer products and services. All of these activities have some impact on the environment. The sheer scale of human technological activity puts an increasing stress on the natural environment to the extent that it cannot absorb our wastes, while our profligate lifestyles lead us to increasingly exploit the planet's limited resources.

Energy resources are an obvious example of limited resources whose use can have major impacts. Figure 1.1 shows the gigantic leap in energy use

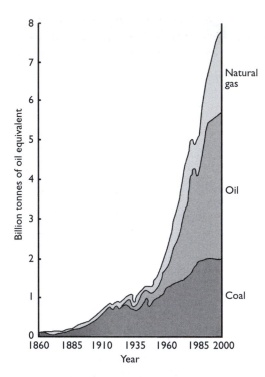

Figure 1.1 _Growth in total fossil fuel consumption worldwide since the industrial revolution (in billion tonnes of oil equivalent)_
(_Source: Physics Review_ 2(5), May 1993, based on data from D.A. Lashof and D.A. Tirpak (eds), _Policy Options for Stabilizing Global Climate_, US Environmental Protection Agency. Draft Report to Congress, Washington, DC, 1989. Reproduced by permission of Philip Allan Updates. Updated from 1985 to 2000 using data from 'Vital statistics', Worldwatch Institute, 2002)

since the industrial revolution. Certainly energy use is now central to most human activities and many of our environmental problems could be described in terms of our energy-getting and energy-using technologies. The most obvious environmental impacts are the physical impacts of mining for coal and drilling for oil and gas, and distributing the resultant fuels to the point of use. However, increasingly it is the use of these fuels that presents the major problems. Burning these fuels in power stations to generate electricity, or in homes to provide heat, or in car engines to provide transport, generates a range of harmful gases and other wastes, and also, inevitably, generates carbon dioxide, a gas which is thought to play a key role in the greenhouse 'global warming' effect. We will be discussing global warming and climate change in detail later, but certainly there seems likely to be a link between the continuing, seemingly inexorable, rise in global carbon dioxide emissions, shown in Figure 1.2, the gradual resultant rise of carbon dioxide levels in the atmosphere, shown in Figure 1.3, and the continuing rise in planetary average surface temperature, shown in Figure 1.4.

If this trend continues, the world climate could be significantly changed, leading, for example, to the melting of the ice caps, serious floods, droughts and storms, all of which could have major impacts on the ecosystem and on human life. Some of the predictions are certainly very worrying. For example, according to the UK Meteorological Office Hadley Centre 'failure to act now on Climate Change could mean the Amazon rainforest is devastated; large sections of the global community

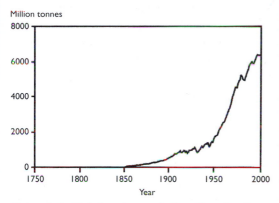

Figure 1.2 *Global carbon emissions from fossil fuel combustion (in millions of tonnes of carbon)* (*Source:* Worldwatch Institute, 'State of the world', copyright 2002, http://www.worldwatch.org)

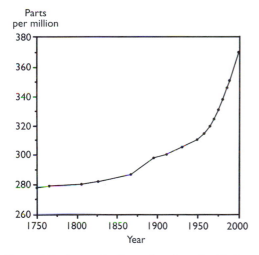

Figure 1.3 *Concentrations of carbon dioxide in the atmosphere (in parts per million)* (*Source:* Worldwatch Institute, 'Slowing global warming: a worldwide strategy', Worldwatch Paper 91, copyright 1989, updated to 2000 from data in 'Reading the weathervane', Worldwatch Paper 160, 2002, http://www.worldwatch.org)

go short of food and water; many heavily populated low-lying coastal areas are flooded and deadly insect-borne diseases such as malaria spread across the world'. However, it is not just a matter of warming, or only a problem for developing countries or tropical areas. There is the possibility that cold water flowing south due to the melting of the polar ice cap could disturb the Gulf Stream which could result in average temperatures in the UK falling by around 10 °C.

Of course there remain many uncertainties over the nature and likely future rate of climate change. Nevertheless, in simple terms it seems obvious that something must change if significant amounts of the carbon dioxide are released into the atmosphere. This gas was absorbed from the primeval carbon dioxide-rich atmosphere and was trapped in underground strata in the form of fossilised plant and animal life. We are now releasing it by extracting and burning fossil fuels. It took millennia to lay down these deposits, but a large proportion of these reserves may well be used up, releasing trapped carbon back into the atmosphere, within a few centuries.

In addition to major global impacts like this, there are a host of other environmental problems associated with energy extraction, production and use – acid emissions from the sulphur content of fossil fuels being just one. Air quality has become an urgent issue in many countries, given the links it has to health. The

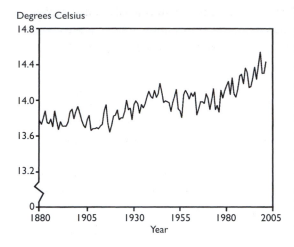

Degrees Celsius

Figure 1.4 *Global average temperature at the earth's surface (in degrees Celsius)*
(*Source:* Worldwatch Institute, 'Reading the weathervane', Worldwatch Paper 160, copyright 2002, http://www.worldwatch.org)

release of radioactive materials from the various stages of the nuclear fuel cycle represents an equally worrying problem, even assuming there are no accidents. Accidents can happen in all industries – and they present a further range of environmental problems, the most familiar being in relation to oil spills.

Ever since mankind started burning wood, air pollution has been a problem, but it became worse as the population increased and was more concentrated in cities, and as industrialisation, based on the use of fossil fuels, expanded. Concern over environmental pollution became particularly strong in the 1960s and 1970s, in part following some spectacular oil spills from tankers, including the *Torrey Canyon* off Cornwall in 1967 and the *Amoco Cadiz* off Brittany in 1979. However, the main concern amongst the emerging environmental movement was over longer term strategic problems: in the mid-1970s, following a series of long-range energy resource predictions (and, as we shall see later, the experience of the 1974 oil crisis), it was felt that some key energy resources might run out, or at least become scarce and expensive in the near future (Meadows *et al.* 1972).

Certainly, a substantial part of the world's fossil fuel reserves have been burnt off and a substantial part of the world's uranium reserves have been used, although, as discussed in Chapter 3, there are disagreements about precisely what the reserves are, and from the 1980s resource scarcity was seen as a less urgent problem. The more important strategic question nowadays is whether what resources are left can be used safely.

Sustainability

The basic issue is one of environmental sustainability: can the planet's ecosystem survive the ever increasing levels of human technological and economic activity? The planetary ecosystem consists of a complex, dynamic, but also sometimes fragile, network of interactions, some of which can be disrupted or even irreversibly damaged by human activities.

In recent years, following the report on *Our Common Future* by the Brundtland Commission on Environment and Development in 1987, and the UN Conference on Environment and Development held in Rio de Janeiro in 1992, the term **sustainable development** has come into widespread use to reflect these concerns. The Brundtland Commission defined it in human terms, as 'development that meets the needs of the present generation without compromising the ability of future generations to meet their own needs' (Brundtland 1987: 43).

In this formulation, the emphasis is mainly on material levels of resource use and on pollution, but the term 'needs' is also wider, and might be thought of as also reflecting concerns about lifestyle and quality of life, as well perhaps as global inequalities and redistribution issues.

Some radical critics of current patterns of energy and resource use go beyond just the issue of environmental impacts, resource scarcity and ecosystem disruption. For some, it is not just a matter of pollution or global warming but also a matter of how human beings live. For them, as well as being environmentally unsustainable, modern industrial technology underpins an unwholesome and unethical approach to life. Technology, at least in the service of modern industrial society, leads, they say, not to social progress but to social divisions, conflicts and alienation, and underpins a rapacious, consumerist society in which materialism dominates. Some therefore call for sustainable alternatives to consumerist society (Trainer 1995). Other critics go even further and, from a radical political perspective, challenge the whole industrial project, and the basic concept of 'development', which they see as, in practice, reinforcing inequalities, marginalising the poor, exploiting the weak and disadvantaging minorities, destroying whole cultures and species, and irreversibly disrupting the ecosystem, all in the name of economic growth for a few.

Even leaving extreme views like this aside, from a number of perspectives, the interaction between technology, the environment and society would seem to be a troubled one. Clearly, it is not possible, in just one book, to try to explore, much less resolve, all these issues. Nevertheless, it may be possible to get a feel for some the key issues and factors involved.

A model of interactions

What follows is a very simplified model of the conflicting interests that exist in society, which may help our discussion, in that they may influence interaction between humanity and the environment. Put very simply, there would seem to be three main human 'domains' that interact on this planet with each other and with the rest of the natural environment. First, there are the *producers* – those engaged in using technology to make things or provide services. Second, there are the *consumers* – those who use the products or services. Third, there are those who own, control, and make money from the process of production and consumption, chiefly these days, *shareowners*. In addition you might add a fourth, meta-group, that is *governments*, nationally and supra-nationally, who, to some extent, 'hold the ring', i.e. they seek to control the activities of the other human groups, for example by developing rules, regulations and legislation.

Obviously, these are not exclusive groups: in reality people have multiple roles. Producers also consume, even if all consumers do not produce. Producers and consumers may also share in the economic benefits of the production–consumption process, e.g. as shareholders. Moreover, there will be differences and conflicts within each of these groups: not all the producers or consumers or shareholders will necessarily have the same vested interests. However, in general terms, the conflicts between and amongst the three groups or roles will probably be larger than those within each group.

In the past, there has always been a conflict between producers and 'owners' – labour versus capital if you like. While the interests of 'capitalists' (i.e. owners or shareholders) are to get more work for less money, battles have been fought by trade unionists to squeeze out more pay. A similar but less politically charged conflict also exists between consumers and capitalists/shareowners. Consumers want good cheap products and services and capitalist/shareowners want profits and dividends. Over the years governments have intervened to control some aspects of both these interactions – for example, to limit the health and safety risks faced by workers and to ensure that certain quality standards are maintained in terms of consumer products and services. In parallel consumers themselves have organised to protect their interests.

In some circumstances the interests of consumers and producers may also clash: consumers want good, safe, cheap products and producers want reasonable pay and job security. The 'capitalists', i.e. the owners and their managerial representatives, tend to have the advantage in most situations: they can set the terms of the conflicts. Thus they may argue that pay rises will, in a competitive consumer market, lead to price increases, reduced economic performance and therefore to job losses, and in general, they can set the terms of employment and of trade. However, they can be constrained by effective trade unions or by market trends created collectively by consumers.

The environmental part of interaction

So much for the human side of the model. The other element is the natural environment: the source of resources from which producers can make goods for consumers and profits for capitalists. The natural environment has no way of responding actively to the human actors, unless you subscribe to the simplified version of the Gaia hypothesis, as originally developed by James Lovelock, which suggests that the planet as a whole has an organic ability to act to protect itself; the various elements of the ecosystem act together to ensure overall ecosystem survival (Lovelock 1979). Nevertheless, even if the natural environment is passive, it represents a constraint on human activities.

Describing this situation more than a century ago, the German philosopher Karl Marx argued that there could in principle be a conflict between the human actors in the system and the natural environment, but that the constraints on resource availability, and the environmental limits on getting access to resources, were far off. He and his followers therefore devoted themselves to the other more immediate conflicts – between the human actors. Nevertheless, it was recognised that at some point, as human economic systems expanded, they would come into serious conflict with the environment.

You could say that this point has now been reached. The rate of economic growth and technological development has brought industrial society to its environmental limits. Some radical critics argue that such is the desire for continued profit by capitalists that, in the face of effective trade unions on one hand, and tight consumer markets on the other, there has been a tendency to increase the rate of exploitation of nature, thus heightening the environmental crisis (O'Connor 1991).

Certainly, mankind has always exploited nature, just as capitalists have exploited producers and consumers, and this process does seem to have increased. However, just as the latter two human groups have fought back, so now the planet is beginning, as it were, to retaliate, by throwing up major environmental problems. Now, as already suggested, apart from putting constraints on some human activities, the natural environment cannot fight back very actively or positively: it requires the help of human actors, environmental pressure groups and governments to protect and promote what they see as its interests.

The model reviewed

The model is now complete: there are the three conflicting human groups (producers, consumers and investors/shareholders) locked into economic conflict; governments active nationally and globally to varying extents; and the natural environment. The environment is mainly dependent for protection on the interventions of people and governments, but perhaps the environment is also able to constrain human activities by, as it were, imposing costs on human activities if these disturb key natural processes, and, in the extreme, making human life on earth unviable.

The issues facing those involved with trying to diffuse this complex situation are many and varied, and, by focusing on economic conflicts, our 'interest' model only partly reflects the political, cultural and ideological complexity of society. In reality people occupy a range of roles: they are not just consumers, producers or shareholders. Moreover, while our model may reflect some of the economic conflicts between various groups within in any one country, it does not reflect the conflicts amongst nations or groups of nations. For example, there are the massive imbalances in wealth and resources amongst the various peoples around the world, and conflicts over ideas about how these imbalances and the inequalities should be dealt with. Moreover, the global nature of industrialisation means that there are global-level problems and conflicts, not least since pollution is no respecter of national sovereignty and national boundaries. These problems seem likely to be further heightened by the rapid expansion of global markets and global corporate economic power, in effect strengthening the power of the 'owner/shareholder' group in the model and increasingly excluding some groups from economic and political involvement and influence. Indeed, it could be argued that we should add this excluded and disenfranchised group to our model – an 'underclass' of the dispossessed poor who, around the world, operate on the margins of society, and often experience the worst of the environmental problems. The world's environmental problems cannot be addressed without considering these wider issues, and our model only partly reflects them.

However, despite these failings, the model does at least provide a framework for discussing some of the key human–environment conflicts. The model abstracts the human element for purpose of analysis, but it should be clear that human beings are not in reality separate from the natural environment. Albert Einstein once said the environment was 'everything except me', but in this book the term 'environment' will mean the entire planetary ecosystem and all that exists in it, including human beings. This definition, with humans as part of the environment, is important, in that too often the environment is seen as something outside of humanity – just as a context for human action. The fact that human beings are part of nature of course makes the problem of solving environmental problems even harder: can the part understand the whole? Will mankind's much vaunted intellectual capacity enable it to rise to the challenge? Or will nature impose its constraints on mankind?

Negotiating conflicts of interest

We will return to some of these questions in Part 4 of this book, but for the moment it seems clear that, assuming that human beings can act usefully to protect the environment, there will be a need to find some way in which the conflicting interests of the four main 'domains' in the model outlined above can be balanced. The battles among the three human elements can no longer be allowed to dominate the political and planetary scene: the fourth element, the environment, must also be considered. Indeed some of the purists amongst the environmental movement would go further: rejecting the idea that minor 'pale green' adjustments and accommodations will suffice, the 'deep greens' would argue, adopting a fundamentalist view, that the interests of the natural world should dominate all others, even to the extent of seriously limiting human activities. This view goes well beyond the idea that human beings have an ethical or moral responsibility for environmental stewardship, a view which is seen as patriarchal or paternalistic. 'Deep ecology' writers like Devall argue that human beings should stop thinking of themselves as being the centre of creation and instead adopt an 'eco-centric' viewpoint (Devall 1988).

Some of the more pessimistic 'deep greens' seem to believe that the planet will never be safe until the impact of human beings on the environment has been returned to the level it was before human civilisation, or at least industrialisation, got going on a significant scale. At times it almost seems as if the ultimate 'deep green' or 'deep ecology' prognosis is that the planet would only be safe without a human presence, and that if mankind does not mend its ways, Gaia will arrange just that.

Social equity

Even leaving aside such ultra-pessimistic views, on the assumption that humankind can respond in time and effectively (a rather big assumption of course), some 'doom' scenarios still have some force. For example, it seems possible that some responses to environmental problems could involve major social dislocations. Those in control might feel it necessary, in order to protect their own interests, to impose socially inequitable solutions, seriously disadvantaging some specific human group or groups.

If this is to be avoided, then some way must be found to combine environmental sustainability and social equity. There will be a need for some sort of accommodation or balance that will not only ensure a more sustainable relationship between humanity and the rest of the ecosystem, but will also attempt to reduce rather than increase social and economic inequalities.

The implication would seem to be that no one human group should be expected to meet all the costs of environmental protection: the costs must be shared. For example, let us assume that consumers want greener products. Industrialists will reply that they can be made available but they will cost more. Similarly workers may press their employers for safer and cleaner production technologies for their own sake but also for the sake of the communities in which they live. They are likely to be told by industrial managers that this would add cost and could lead to lower wages or even job losses. What is missing in this formulation is the interests of shareholders. After all, you could argue that everyone should carry the burden: consumers, producers and shareholders.

In reality of course, shifting to greener products and cleaner production processes may not in the end cost more, or at least it may not be a bad move, in the longer term, commercially. As pressures for a 'clean up' grow, for example through government legislation on environmental protection or consumer pressure for greener products, companies that take the initiative will have a competitive advantage compared to those that do not. Of course, if left just to market competition, that means there will be some commercial losers and overall there may well be short-term costs and dislocations in terms of employment and profits. That is precisely why a proper negotiation process is so important – to reach some sort of agreement on the distribution of costs and benefits amongst all the human stakeholders, in the wider context of overall environmental protection.

Clearly, it will not be easy to establish this sort of negotiation process, even if we stay within the confines of our model and apply it just to one country. Quite apart from the powerful vested interests of the competing groups, regulations take time to have an impact, and there are usually ways of avoiding pressures for change, at least in the short term. And once we step outside of a single country focus, the opportunities for evasion become even greater. For example, companies can move to countries where regulatory pressures are weak. So to be effective, an attempt has to be made to extend the framework of negotiation and regulation to all levels, local, national and global.

This process of negotiation of interests, at whatever level, will depend to some extent on actions taken by governments, for example by setting new environmental standards and regulations, although, equally, all the human actors in the system can also play a part, by including environmental considerations in their otherwise partisan negotiations. Despite the difficulties, there have already been attempts to do this, for example in relation to technological development choices, at all levels around the world.

We will look at some examples, and at the concept of social negotiation generally, in Part 4. For the moment, however, we will turn to exploring some of the environmental problems the world faces in more detail and look at some of the potential technological solutions and their limitations. For, as will become apparent, purely technical solutions may not suffice if the aim is to develop a genuinely sustainable and equitable future: social and economic changes and adjustments may also be necessary.

The growth of environmental concern

The social and political dimensions of the problem of devising a sustainable future may become clearer if we look back to the beginnings of the contemporary environmental debate. In the late 1960s and early 1970s there was a perhaps unique concurrence of ideas from a number of social and political movements.

The early 1960s had seen the beginnings of environmental concern, symbolised by the publication in 1962 of Rachel Carson's *Silent Spring* which, amongst other things, warned of the ecological dangers of pesticides like DDT.

Subsequently there was a growth in environmental concern amongst young people, many of whom formed part of a counter-culture, which flowered briefly in the late 1960s and early 1970s. The young people involved were often from relatively affluent backgrounds, but they challenged the ideas of the conventional consumerist and materialist society in which they had grown up. Some were not content with simply objecting to the way things were done at present, but also wanted to create alternatives: alternative lifestyles and alternative technologies to support them. There were self-help experiments in rural retreats with windmills, solar collectors and so on, with their decentralist, communitarian philosophy being underpinned by books like Fritz Schumacher's seminal *Small Is Beautiful* (Schumacher 1973), which argued for the use of smaller, human-scale technology supporting and reflecting a more humane and caring society.

In parallel, the late 1960s and early 1970s saw a rise in radical politics, reflected most visibly by the student 'protest' movements around the world. The 'new left', which emerged as one of the many strands in this movement, challenged the political dogma of the traditional left. The latter held that capitalism, with its single-minded concern for the economic interests of those who owned and controlled technology, denied society as a whole the full benefits of technology; but once freed from capitalist constraints the same technology could be used to meet human needs more effectively. The classic interpretation of this view in practice had been fifty years earlier, when, following the Russian Revolution, Lenin adopted Western production technology and Western technology generally, since he argued that it was the best available at that historical stage. The theoretical point was that the existing technology could simply be redirected to meet new ends. Eighty years on, the shortcomings of this view have become clear: industrial development during the Soviet period seems to have replicated many of the worst examples of environmentally destructive Western technology.

The 'new left' in the late 1960s and early 1970s to some extent foresaw this problem. They argued that technical means and political ends inevitably interacted: old means could not be used to attain the new ends and a new set of technologies was required. Like many environmentalists and the 'alternativists' in the counter-culture, they were arguing for an alternative technology.

Alternative technology

This line of argument had been put forward by a British writer, David Dickson, in the seminal paperback *Alternative Technology and the Politics of Technical Change* (1974). However, Dickson, along with many members of the counter-culture, also felt that a simple switch of technology would not be sufficient: technology and society interacted, so there was a need for an alternative society as a base for the alternative technology. As Dickson put it, 'A genuine alternative technology can only be developed – at least on any significant scale – within the framework of an alternative society' (Dickson 1974: 13). Thus the 'soft' and 'hard' paths to a sustainable future outlined by influential US energy activist Amory Lovins, could not just be defined by 'soft' and 'hard' technology (for example, solar and nuclear respectively), but also required different social and economic arrangements (Lovins 1977).

Indeed, some, following Dickson, felt that society determined technology, so you would need social change first, although a less deterministic, two-way interaction, was usually seen as a more realistic model, implying linked changes in both society and technology (Elliott and Elliott 1976).

With this in mind, rather than a comprehensive confrontational approach, some adopted a more strategic approach to social and technological change, operating in pathfinder mode, initially on the fringe. For example, Peter Harper, an English enthusiast for what he called 'radical technology', claimed that: 'premature attempts to create alternative social, economic and technical organisation for production can contribute in a significant way to the achievement of political conditions that will finally allow them to be fully implemented' (Harper 1974: 36).

Others again warned that any attempts to introduce radical alternatives would be co-opted by commercial interests, and would be shorn of their values and their radical political edge. The emphasis could thus be on selling 'technical fixes', that is just the hardware, and not on implementing the social changes with which alternative technology was meant to be associated. Thus, for example, solar collectors could become just conventional consumer products rather than harbingers of social transformation.

Of course, some would say this did not matter. Certainly, an interest in 'alternatives' does not always necessarily imply or require a collective commitment to 'progressive' social policies of the sort espoused by liberal environmentalists or radicals on the left of the political spectrum. Some people have sought simply to be free of social norms and state control on an individual basis. For some, alternative technologies simply provide a way of escaping from society. In the USA this libertarian response has sometimes shaded into the militantly independent 'survivalist' ideology, often involving extreme right-wing views.

These sorts of developments highlight the weakness in the simple proposition that alternative technologies are automatically linked to progressive social change.

The current 'green' debate

These debates are relevant to our contemporary situation in that the issues are now much clearer. Some alternative technology has been co-opted, some green products have just become luxury items for the well heeled, and some technical fixes are being offered as solutions to our environmental problems, while at the same time some radicals in the contemporary 'green' movement are still arguing that only a radical transformation of society will be sufficient. More optimistically, some argue that alternative technology has simply moved out of a counter-cultural ghetto, beyond a marginal niche market, and into the mainstream (Smith 2002).

Inevitably, there is a wide range of views on these issues in the 'green' movement. This is hardly surprising since the green movement, which emerged in the 1980s, has a multitude of strands. It is much more than just the members of green political parties or activist groups, even though some of the latter are now quite large: tens of thousands of people belong to organisations like Friends of the Earth and Greenpeace. It might also be thought to cover anyone who has some concern for the environment, as reflected, for example, in their support for wildlife protection or in their consumer behaviour.

All these levels of involvement can have an impact, although equally there can also be tactical and strategic divergences and disagreements. Certainly, the growth of consumer awareness has led to pressure for environmentally friendly products and, in turn, for environmentally sound production processes: greener products and cleaner production technologies, which have fewer impacts and use less energy. There is considerable activity in this field at present, but equally there are those who would ask whether this is enough to achieve real sustainability. For example, it has been argued (by Chris Ryan, a leading Australian exponent of ecodesign) that if the various global environmental problems are to be properly addressed, pollution levels and global energy and material resource use must be cut by around 95 per cent, but that this may not be possible just by 'technical fixes'. There may also be a need for social change, for example, in qualitative patterns and quantitative levels of consumption (Ryan 1994).

For radical 'greens' the real issue is this one: can and should growth in material and energy consumption be continued, stimulated by ever growing expectations concerning living standards? Is there not a need for more radical changes – in society as well as technology? Increasingly it is argued that there is a need for a more radical transformation of technology and also possibly of society – an alternative set of technologies better matched to environmental protection, linked to an alternative set of social and cultural perspectives and structures.

Rather than explore these large issues in the abstract, this book attempts to look at the case for and the implications of this type of change by focusing on energy technology as an example. Clearly, this is just one area of technology, but as has been indicated, it is perhaps the central one in environmental terms. The next chapter provides a basic introduction to energy issues, reviewing energy use past, present and future.

Summary points

- Energy use has increased dramatically in recent years and this has had increasingly adverse impacts on the environment.
- The concept of sustainable development has been proposed, which implies that alternative technologies may be needed which avoid or reduce adverse impacts.
- Although alternative technologies may help, there may also be a need for major social changes if sustainability is to be achieved.
- All the various groups in society (e.g. owners/investors, producers, consumers) must share the burden of protecting the ecosystem of which they are all inevitably part.
- Energy use is probably one of the key factors influencing environmental impacts.

Further reading

A useful general overview of the way technology has shaped human–environmental interactions is provided by David Kemp's *Global Environmental Issues* (Routledge, London, 1994). For a good introduction to the history of green thinking see David Pepper's seminal *The Roots of Modern Environmentalism* (Croom Helm, London, 1984) and Andy Dobson's *Green Political Thought* (Routledge, London, 1995).

You might also like to try to track down David Dickson's classic text *Alternative Technology: the Politics of Technical Change* (Fontana, London, 1974). Although this is out of print, it should still be available in libraries. The various strands of thinking that make up the 'deep green' or 'deep ecology' approach are explored in Bill Devall and George Sessions' classic text *Deep Ecology* (Peregrine Smith Books, Salt Lake City, 1985). Sessions' most recent book, *Deep Ecology for the Twenty-First Century*, is available from Schumacher Books, Foxhole, Dartington TQ9 6EB, UK, along with several other similar titles.

For a somewhat more conventional viewpoint on environmental problems and policies see the classic report *Our Common Future* by the Brundtland Commission on Environment and Development, which was published by Oxford University Press in 1987.

Moving up to the current debate on the environment, there is a vast and growing literature on sustainable development. For a good general introduction to the issues see, for example, Paul Ekins, *Economic Growth and Environmental Sustainability* (Routledge, London, 1999).

Finally, for an extensive and critical review of green thinking, see Bjorn Lomborg's *The Skeptical Environmentalist* (Cambridge University Press, Cambridge, 2001).

 2 Energy and environment

- Energy units
- Acid rain
- Global warming
- Nuclear opposition

In order to explore the environmental implications of energy generation and use, we will need an understanding of the basic terms, concepts and measurement units used in the energy field. As you will discover, perhaps surprisingly, some of the measurement units are far from uncontroversial. Having established this context, the chapter then looks at how energy is and has been used around the world, and at some of the social and environmental problems that have emerged as a consequence of the use of fossil and nuclear fuels. Finally we look briefly at what the alternative energy options might be.

Energy and its use

Energy is a concept rather than an actual thing: we say people have energy when they can work or play hard. The manifestation of energy in material terms is 'fuel', and these two terms, energy and fuel, tend to be used interchangeably. The concept of power is also often used as if it meant the same as energy.

To set the scene we need to have a clearer understanding of some of the basic units and terms used in the discussion of energy issues.

Although it is common to talk of 'energy generation' and 'energy consumption', strictly, energy is never 'created' or 'consumed', it is just 'converted' from one form to another. The term *power* is used to describe the conversion capacity of any specific device, i.e. the rate at which it can convert energy from one form to another, and the unit most commonly used is the *watt*. Strictly, it is a measure of the 'capacity to do work'.

Specific energy 'generating' or 'consuming' devices are therefore given a *power rating* (or *rated capacity*) in watts and multiples of watts, e.g. a *kilowatt* (kW) is 1,000 watts. The *megawatt* (MW) is 1,000 kilowatts (or 10^6 watts), the *gigawatt* (GW) is 1,000 MW (or 10^9 watts), and the *terawatt* (TW) is 1,000 GW (or 10^{12} watts). To give you an idea of scale, a typical large modern coal or nuclear power station has a rated capacity of around 1.3 gigawatts (GW), while in the mid-1990s the UK had around a total of 65 GW of electricity 'generating' capacity.

The amount of *energy* converted ('generated' or 'consumed') or more accurately, the actual 'work' done, is defined by the power of the device multiplied by the time for which it is used (i.e. watts × hours). It is usually measured in *kilowatt hours* (kWh). This is the unit by which electricity and gas is sold in many countries (although, obscurely, the USA still makes use of British thermal units, the old measure for the heat content of fuels: 1 kWh = 3413 BTUs). A typical 1 kW-rated one-bar domestic electric fire 'consumes' 1 kilowatt hour (kWh) each hour.

For larger quantities, multiples of kWhs are used, most commonly the *terawatt hour* (TWh) which is 1,000,000,000 kWh or 10^9 kWh. To give an idea of scale, at the turn of the century, the total UK *electricity* 'consumption' was about 370 TWh per annum. Remember, however, that this is the figure for the 'consumption' of electricity, not total energy consumption: it does not include all the direct *heat* supplies. For example, domestic consumers in the UK used around 370 TWh worth of energy from natural *gas* for heating in 2000. In addition there are the fuels used for *transport* and in *industry*.

Rather than focusing on figures for energy consumption in its various forms at the point of use, the total amount of energy used is often measured in terms of **primary energy** consumption, that is the amount of energy in the basic fuels used by energy conversion devices, whether for electricity production, heating or transport.

However, it is important to remember that 'primary energy' figures, for the total energy in the fuels used by energy conversion devices, are much larger than the *finally delivered* energy, as utilised by consumers, since there are losses in the conversion process in power plants and in transmission to users. This is particularly true of electricity: conventional coal- or nuclear-fired power plants only have conversion efficiencies of around 35 per cent. Even the best modern combined-cycle, gas-fired power stations can only convert around 50 per cent of the energy in the input fuel into electricity. Moreover, after it has been produced, up to 10 per cent of the electricity may be lost when it is transmitted along power lines to consumers, depending on the distances involved. Finally, consumers will use this 'delivered' energy to power a variety of energy conversion devices with varying degrees of efficiency, with much of it often being wasted, for example, in poorly insulated buildings. Primary energy figures therefore only tell part of the story. As we shall see in subsequent chapters, there is also a need, when comparing technologies and energy systems, to consider the overall efficiency of energy conversion and transmission, and the use to which the energy is put.

The battle of the units

As can be seen from the previous discussion, measuring energy use is not as simple as it might seem. Given that there are many ways in which energy is generated and used, it is not surprising that there are many different, often confusing, ways in which it is measured and many devotees of rival systems of measurement. We have mentioned kWh, which is the most familiar unit to most people since it is what is used on consumers' bills. However, energy analysts sometimes use the basic physical unit for 'work', the *joule* (J) or multiples of joules. One watt is one joule per second, so a kWh is 3,600,000 joules, and the joule is thus a very small unit. Hence large multiples are common, e.g. peta-joules or PJ (1,000 tera-joules) and exa-joules or EJ (1,000 peta-joules).

In the UK until recently, for statistical comparison purposes, primary energy use was also often measured in terms of the equivalent amount of coal that would be required to be burnt to provide that energy regardless of what fuel was actually used in power stations, i.e. in 'tonnes of coal equivalent' (or more usually 'million tonnes of coal equivalent' or 'mtce'). This no doubt reflected the historical predominance of coal in the UK's economy. In 1994, the UK government's statisticians decided to adopt the European standard unit, with the energy content of all fuels being rendered, for statistical comparison purposes, in terms of the equivalent amount of *oil* that would have the same amount of energy content. The energy content of all fuels is therefore now presented in terms of *tonnes of oil equivalent*, or more usually, *million tonnes of oil equivalent* (mtoes). However, for the purposes of this book we will stay with the more familiar kWh, TWh, etc. figures. For reference 1 mtoe = 11.63 TWh, and 1 TWh = 0.086 mtoe.

Interestingly though, as concerns about climate change have grown, the literature on energy policy has increasingly been filled not with data presented in terms of kWh, TWh or million tonnes of oil equivalent, but with measure of the tonnages of carbon dioxide emissions – usually conveyed in terms of 'millions of tonnes of carbon dioxide'. However, sadly, there is a further complication. As we will see later in this chapter, carbon dioxide is only one of the gases thought to be responsible for altering the way solar radiation interacts with the upper atmosphere – creating the so-called 'greenhouse' effect, which is said to be the basis of global warming and climate change. There are several other gases involved, including methane. Although the volumes involved are less, methane actually has much more of a 'greenhouse' effect, molecule for molecule, than carbon dioxide, by a factor of 21. To provide a way to compare overall greenhouse gas emissions on the same basis, the emission figures for all the greenhouse gases from any particular energy technology are therefore converted into the equivalent tonnage of carbon dioxide that would have the same effect. So the tonnage of methane is converted into the equivalent tonnage of carbon dioxide and is added to the tonnage of actual carbon dioxide. The resultant figure is then presented as tonnes of carbon *equivalent*, or, more usually, millions of tonnes of carbon equivalent. Note that it is 'carbon' here not 'carbon dioxide'. This is because, just to make it all more difficult, the figure is converted to just the tonnage of *carbon*, rather than carbon dioxide, which involves a reduction by a factor of 0.27 (which is the ratio of the molecule weights of carbon and carbon dioxide). To convert back to carbon dioxide, multiply the figure for carbon by 3.7.

This may all sound rather complex, but you need to watch out for how these units are rendered. And it is perhaps interesting that, rather symbolically, we have moved over in the last few decades from an emphasis in energy statistics on 'million tonnes of coal equivalent' (a measure of fuel obtained from the ground) to 'million tonnes of carbon equivalent' (a measure of the gas released into the atmosphere).

One final warning, when looking at official statistics on energy use, for example for specific countries, the figures for energy production will usually differ from the figure for energy consumption, since (quite apart from the usual losses in conversion and transmission) some energy may be exported, some may be imported and some may be stored.

Not everyone will find the details of the units used in energy measurement exciting, but obviously it is necessary to have a common measure, and establishing this may not always be as uncontentious as it may seem. In this regard, before we move on to look at the way energy is actually used, it may be worth noting an extra complication with how it is measured.

In addition to switching to the use of 'million tonnes of oil equivalent', the changes introduced in 1994 to the way energy statistics are rendered in the UK also involved a subtle shift in the way the energy content is calculated. It is now based on the energy content of the output power produced by power plants, rather than on the energy content of the input fuel needed to generate the power. This has some interesting results. Previously, on the so-called 'fuel substitution' basis, the energy contribution from non-fossil fuel-powered devices like the wind turbines was calculated in terms of the energy content of the fossil fuel that would have to be fed to a conventional power station to provide the same amount of power output. On the new 'energy supply' basis, the figures for contributions from devices like wind turbines drop dramatically, by 73 per cent, this being the scale of the losses associated with energy conversion in conventional power plants. The end result of the change was that the contribution in 1993 from the so-called renewable energy sources (hydroelectricity, wind power, etc.) became 0.4 mtoe instead of 1.4 mtoe. (Department of Trade and Industry 1994a).

Not surprisingly, the new approach is not particularly popular with renewable energy supporters. Interestingly the figures for nuclear power plants are not much affected by this new way of rendering energy statistics. Fortunately though, the figures are also made available on the original basis, since it was accepted that, while it was useful to be able to compare the actual amount of power being supplied to the grid, it was also useful to be able to assess the degree to which renewable sources were substituting for fossil fuel. But this methodological aside should at least alert you to the need to be careful about the way units are used.

National and global energy use

Having now established some of the basic energy units, we can move on to look briefly at how energy is actually used around the world. Primary energy figures can be derived at various level – for countries, or for the world as a whole. Within the national context, primary energy use is often broken down in terms of its type.

Clearly, the mix of sources used will vary around the world. For example, Norway and Brazil obtain more than half their electricity from hydro, and France obtains more than 70 per cent of its electricity from nuclear power, whereas in some developing countries there is very little use of electricity from any source. Currently, the UK gets around 25 per cent of its electricity from nuclear power plants, nearly 30 per cent from coal-fired plants and nearly 40 per cent from gas-fired plants. Hydro and new renewables produce around 3 per cent of the UK's electricity.

In overall terms, the industrialised countries, with about 20 per cent of the world population, use about 60 per cent of the world's energy, and that translates into an even larger imbalance in per capita terms – the richest billion people use five times more energy than the poorest 2 billion. However, some parts of the developing world are beginning to catch up in terms of total national energy consumed per annum.

The pattern of energy production has undergone some significant changes in recent years, particularly in relation to electricity generation in some of the industrialised countries, with gas beginning to replace coal. Thus in 1992, 60 per cent of the UK's electricity came from coal-fired power stations, 21 per cent from nuclear plants, 8 per cent from oil-fired plants, 2 per cent from hydroelectric plants and 4 per cent from gas and other fuels, with 5 per cent being imported (Department of Trade and Industry 1993). But by 2000, only 28 per cent of the UK's electricity came from burning coal, while gas had a 39 per cent share, primarily due to the use of natural gas fuelled turbines for electricity production (Department of Trade and Industry 2000).

Similar trends are occurring elsewhere in the world, with gas becoming a major new source of electricity as well as heat. Table 2.1 shows the global primary energy use in 2000, by source.

Whereas the pattern of supply and demand within individual countries can, as we have seen, vary significantly from year to year, the global percentages, which aggregate the patterns of a large number of countries, do not tend to change rapidly. However, the overall trend globally is upwards, with the growth rate being typically 1–2 per cent per annum.

Table 2.1 *Global primary energy consumption in 2000, by source (%)*

Hydro	2.3
Nuclear	6.6
Oil	34.6
Biomass	11.3
Gas	21.4
Coal	21.6
New renewables	2.1

Source: World Energy Council. See their web-site for regular updates: http://www.worldenergy.org

Looking further into the future, given the growing world population and rising material expectations, energy use globally is likely to continue to increase. Later in this book, we will be looking at a range of energy scenarios that have been devised to attempt to map out possible patterns of longer term development in energy supply and demand. Some assume that energy demand is likely to increase up to perhaps three times current levels by the year 2060.

However, predicting the future is difficult: past trends are not necessarily a good guide to what might happen next in terms of fuel availability and prices, or in terms of patterns of energy consumption in the various rapidly changing sectors of energy use around the world. The future is far from being predetermined: part of the aim of this book is to explore what choices there are in terms of creating new patterns of supply and demand.

The growth of energy use

Having now set the statistical scene in terms of general patterns of energy use, let us now look in more detail at how this pattern of energy use emerged historically. For, although past trends may be a poor guide to future patterns of development, in order to begin to think about the future, we need to look at how the present pattern of energy use came about.

In ancient times and right up to the beginning of the industrial revolution, motive power was provided either by direct human effort, including the use of slaves, or by animals or, at sea, by wind. Wood was used for heating, cooking and some processing of materials. In the Middle Ages, machines such as windmills and watermills gradually took on some of the load, and watermills actually played a key role in the early stages of the industrial revolution. However, the industrial revolution only really got going when coal began to be used widely. Whereas, before, factories had to be sited near power sources such as rivers, and also possibly near sources of raw materials, coal could be transported to factories – in trains powered by coal-fuelled steam engines – to run the factory equipment. Coal also provided a fuel for trains and ships to shift raw materials from remote sites to factories, and to transport their finished products to distant markets. Moreover, coal-fired engines could be used to pump water out of mines, thus allowing deeper mines to be used, and more coal to be produced. The combination of coal, steam engine and train technology was thus a real industrial and economic breakthrough.

The role of coal as the key fuel was not challenged for hundreds of years: coal was used for heating houses and was also converted into gas for use in lighting and heating. It powered trains and ships and it was fed to power stations to generate electricity. In some countries use could also be made of hydroelectricity from large dams, but in general the first main rival to coal was oil. This found a use in some industrial processes and, in the form of petrol (petroleum spirit extracted from oil) it became the main fuel for vehicles. Initially, in the 1890s, most of the first automobiles were powered by electricity or steam, produced using coal, but by the beginning of the twentieth century the petrol-powered internal combustion engine had taken over. Similarly for shipping – oil replaced coal.

In the inter-war years, while electricity became increasingly important in many sectors of the economy, oil and gas also gradually took on increasing roles. In the years after the Second World War, oil began a steady rise and by the time of the first oil crisis in 1973–4, its use, at least in the advanced industrialised countries, had overtaken coal. At this point a new direct source of gas (natural methane) was also found – from under the seabed – and 'natural gas' rapidly replaced the so-called 'town gas' that had previously been produced from coal. In parallel, following the war effort to produce an atomic bomb, civil nuclear power had begun to make a small contribution to electricity supplies in most of the developed countries.

So as the modern industrialisation process got underway in the advanced Western countries in the period after the Second World War, the basic energy pattern was fairly stable: roughly speaking, coal, oil and gas had very roughly equal shares in primary energy use terms, but with oil beginning to dominate, and with nuclear power (along with hydroelectric plants) making a small additional contribution. Basically, coal provided the bulk of electricity, gas provided the bulk of heating, oil the bulk of transport fuel. The exact proportions differed from country to country, depending on the availability of indigenous reserves. For example the UK had ample coal, as did the USA. In addition the USA had oil, although they also imported it from the Middle East, as did the UK and many other industrialised countries. By contrast, some of the newly emerging countries like Japan had few coal reserves and had to rely heavily on imported oil.

The oil crisis

The 1973–4 oil crisis was precipitated by the Yom Kippur War between Israel and the Arab states. OPEC, the Organisation of Petroleum Exporting Countries, which at that point was dominated by the Arab states, objected to the West's support of Israel and imposed a sudden increase in oil prices. This had a dramatic effect on the industrialised economies: they suddenly realised how reliant they had become on imported oil. Over the next decade strenuous technical efforts were made both to find other sources of oil and to diversify away from oil, although equally strenuous political, and in some cases military, efforts were also made to secure access to Middle East oil.

The oil crisis coincided with the first serious signs of concern about the environmental impacts of the use of fossil fuel, although the main concern, as highlighted graphically by the oil crisis, was about fuel scarcity. As a result of what was seen as an 'energy crisis', a range of new technologies were looked at – including the use of natural energy sources like the winds and solar heat, which are collectively called **renewable energy** sources, since, unlike the fossil fuels, they will never be exhausted. Nuclear power, which had been developing relatively slowly in the West since the war, was also given an extra boost.

By the early 1980s, however, the pattern had settled back down: oil prices had dropped in real terms, natural gas had become a new and cheap fuel. Nuclear power had expanded, but not as dramatically as its proponents would have liked. It was bedevilled by economic and technical problems. The accident at Three Mile Island in Pennsylvania in 1979 effectively halted progress in the USA, and worldwide the total nuclear contribution hovered around 5 per cent of primary energy, reaching 5.7 per cent by 1990. Renewable energy technology was still in its infancy, although it is important to remember that, worldwide, hydroelectric dams generate around 20 per cent of the world's electricity from a renewable resource, and in some countries (notably Brazil and Norway) hydro power provides the largest single energy input. Overall though, in most industrial countries, coal, oil and, increasingly, gas remained the dominant options.

Environmental problems emerge

Although, as noted earlier, environmental issues became prominent in the early 1970s, some of the key issues had emerged earlier. For example, environmental pollution from coal burning had emerged as a key issue in the UK as a result of a series of disastrous 'smogs' in the 1950s, when there were many deaths. One smog in London in December 1952 led to an estimated 4,000 deaths in the following weeks. One response was the 1956 Clean Air Act, which limited the use of coal in open grates in urban areas. An attempt was also made to disperse emissions from power plants and industrial chimneys by increasing the heights of the chimney stacks. This technical fix worked up to a point – although the result was that some of the key pollutants were carried over longer distances and began to have an impact far away, for example on the Scandinavian ecosystem.

Similar problems existed with emissions from power plants elsewhere in the world with, for example, Germany's Black Forest also suffering damage. By the 1980s the problem could no longer be ignored. As a result of objections by the Scandinavians and the campaign work of environmental groups, there was a renewal of environmental concern over the impacts of the use of fossil fuel. By 1983 Germany had embarked on a programme of removing sulphur dioxide gas from power station emissions, although some other countries, like the UK, took longer to react.

The key issue was that of **acid rain** – the consequence of acid emissions from power stations, caused primarily by the sulphur content of coal and oil. When burnt, the small sulphur element in these fossil fuels is converted into sulphur dioxide gas, which can dissolve in water to produce weak sulphuric acid. This, precipitated in the rain, can damage trees, and, collected up in lakes, can injure wildlife and fish. Acid rain also damages buildings and crops. Oxides of nitrogen can also be produced by the combustion of fossil fuels in power stations, and also from the combustion of petrol in cars, and they can play a similar role.

The acid emissions from the combustion of fossil fuels can be filtered out relatively easily, and, as noted above, although it adds to the cost, this is now done quite widely. By contrast it is less easy to see how to deal with the next problem that emerged, that is the emission of greenhouse gases such as carbon dioxide. Carbon dioxide is the fundamental product of combustion: put simply burning means using oxygen from the air to convert the carbon in fossil fuels into carbon dioxide gas plus heat.

Fossil fuels are hydrocarbons: they contain varying amounts of hydrogen and carbon. All fossil fuels create carbon dioxide when burnt, but some produce more than others depending on their chemical make up. Coal has

Table 2.2 *Carbon dioxide production*

Fuel	kg of CO_2 per GJ of heat
Coal	120
Oil	75
Natural gas	50
Biomass	77

a high pure carbon content, and when burnt it therefore produces mostly just carbon dioxide and heat. By contrast methane (natural gas) is made up of one atom of carbon plus four of hydrogen and when burnt the hydrogen is converted into water, so that the ratio of carbon dioxide to heat produced differs. The result is that the combustion of natural gas produces less carbon dioxide per unit of energy generated than the combustion of coal. By contrast the combustion of the more complex hydrocarbons, like oil, produces intermediate levels of carbon dioxide. See Table 2.2.

Carbon dioxide is what is in fizzy drinks: it is a tasteless, colourless and chemically inactive gas which dissolves only very slightly in water. That makes it hard to remove from power station or car emissions. In theory you could freeze it out from power station emissions, producing solid carbon dioxide or 'dry ice' – at vast expense. Alternatively, as we will be discussing in Chapter 4, you could collect and store the gas – again at significant cost. The most direct way to avoid carbon dioxide levels building up is not to burn fossil fuels.

Global warming

Global warming is of course the reason why there is a need to avoid producing carbon dioxide. Gases like carbon dioxide travel up into the upper atmosphere (the troposphere) where they act as a screen to sunlight. They allow the sun's rays in but stop the heat radiation from re-emerging, much as happens with the glass in a greenhouse. The result is that the greenhouse, in this case the whole world, heats up. Some degree of global warming via this type of greenhouse effect is actually vital, otherwise this planet would be too cold to support life. However, the vast tonnage of carbon dioxide gas we have released into the atmosphere seems likely to upset the natural balance.

The carbon dioxide level in the upper atmosphere reached 368 parts per million (ppm) in 2000, compared with the pre-industrial figure of 280 ppm and the projections are that on current trends it could reach at least double pre-industrial levels by 2050, and, unless emissions were reduced, could reach 1,000 ppm by 2100. On the basis of current trends, world

CO_2 emissions are projected to rise from 6.1 billion metric tonnes of carbon equivalent in 1999 to 9.8 billion metric tonnes in 2020, a 60 per cent increase, possibly rising to 15.1 billion metric tonnes by 2050, a 152 per cent increase on 1990 levels (IEA 2001a).

Even if we stopped all emissions immediately, it would take time before the situation stabilised, since the carbon dioxide imbalance that has already been created is likely to persist for some while; although there are continual interchanges between the air, the sea and the other carbon 'sinks', the net excess carbon dioxide can take many decades, and possibly longer, to be absorbed.

The situation has been made worse by the fact that large areas of forest around the world have been felled, thus removing an important 'sink' for carbon dioxide, since trees absorb it as they grow. Re-afforestation is obviously vital, since amongst other things it helps maintain biodiversity, but it would take many decades just to replace what has been lost. Unless carried out on a vast scale, re-afforestation and the creation of new forest could only play a small role in absorbing the vast amount of carbon dioxide released every year by power plants and cars, quite apart from dealing with the emissions that have already been made. For example, the afforestation of a land area as large as Europe from the Atlantic to the Urals would be required in order to 'sequester' or absorb the amount of carbon estimated to be emitted from the burning of fossil fuels during the first half of the twenty-first century. Moreover, it would not be permanent storage. Trees eventually die, and can also catch fire, thus re-emitting the stored carbon dioxide. So the forests would have to be continually renewed. It might be possible to establish new forests in some parts of the world, where there is less pressure on land, but in the UK context, to sequester all the UK's continuing carbon emissions in trees would require new forests to be planted over an area the size of Devon and Cornwall *every year*.

Carbon dioxide is not the only culprit. Methane gas, produced naturally from marshes, cows and other ruminating animals and also from human wastes, plays a very significant role in global warming, much more, molecule for molecule, than that played by carbon dioxide. Ozone gas (O_3) also plays a part in the complex chemical interactions in the troposphere affecting the way the sun's radiation is absorbed.

However, it is worth noting at this point that the greenhouse effect is unrelated to the creation of so-called 'ozone holes' in the stratosphere. Although ozone depletion plays a role in the greenhouse effect, the ozone holes are a separate phenomena – the result of a chemical interaction between CFC (chlorofluorocarbon) molecules, which destroy ozone, notably in polar regions. CFCs are man-made chemicals which were developed as a supposedly inert gas for use in refrigerators and foam packaging, amongst other things. However, it turns out that CFCs not only lead to ozone holes but are also powerful greenhouse gases.

The results of stratospheric ozone depletion due to the release of CFCs are that dangerous wavelengths of solar radiation can reach the earth's surface where they can cause cancers and damage plant growth.

The results of global warming, if it happens on a significant scale, are likely to be even more severe. It is well to remember that the ice age, millennia ago, only involved a global temperature variation of around 4 °C. The global warming that seems to be occurring now could involve larger changes in temperature. One result could be that ice-caps would melt. Given that most of the Arctic ice is floating, this would not lead to sea-level rises, but much of the Antarctic ice sheet is on rock and the melting of this, coupled with the effects of the thermal expansion of the seas, would give rise to significant sea-level rises. Global warming could also lead to the disruption of crop growing as climate patterns change. It would not be simply a matter of increased temperatures: the climate system would become more erratic, with more storms and more droughts – hence the use of the more general phrase *climate change*. Flooding could be very severe in some parts of the world, particularly in low-lying areas such as the Netherlands and Bangladesh, with the risk of great loss of life and the loss of areas for living and growing food. Some low-lying islands, for example in the Pacific, could disappear entirely. See Box 2.1.

Given that the impact of global warming on life on earth could be very dramatic, insurance companies around the world are already taking the issue seriously. Certainly the last decade has seen some dramatic changes in weather and climate patterns, and the consensus is increasingly that climate change is a reality. In addition, the scientific consensus is now that this is the result primarily of human activity, with the Intergovernmental Panel on Climate Change (IPCC) suggesting that, on the basis of the most up-to-date models, average global temperatures are likely to rise by between 1.4 and 5.8 °C over the current century. This might lead to average sea-level rises of between 0.09 m and 0.88 m by 2100 (IPCC 2001). See Figure 2.1. The ranges reflect variations in the assumptions used for the climate models, not least over the level of emissions that we will make.

Box 2.1

The impacts of climate change

According to a report by the UK Meteorological Office Hadley Centre, published in 1999, by the 2080s, if action is not taken to tackle climate change:

- large parts of northern South America and central southern Africa could lose their tropical forests;
- 3 billion people could suffer increased water stress – Northern Africa, the Middle East and the Indian subcontinent will be worst affected;
- around 80 million extra people could be flooded each year due to rising sea levels: Southern Asia, South East Asia and island states in the Caribbean, the Indian Ocean and the Pacific Ocean will be most at risk;
- about 290 million extra people could be at risk of malaria – China and central Asia will be most affected;
- the risk of hunger in Africa will increase due to reduced cereal yields.

On a more parochial note, there is also the possibility, mentioned earlier, that cold fresh water flowing south due to the melting of the Arctic ice-caps could disturb the Gulf Stream, which could result in average temperatures in the UK falling by around 10 °C. Fortunately, a complete switch off of the Gulf Stream is currently seen as unlikely in the short to medium term, and, even if it did occur, some of the resultant cooling might be offset by the parallel warming trend.

Although it is hard to predict exactly what the specific local outcomes of changes like this would be, it is clear that not all the impacts would necessarily be negative. For example, in some areas increased temperatures might improve agricultural productivity. But, in general, any local benefits that might follow from climate change seem likely to be somewhat marginal compared with the potential for negative impacts. For example, flooding could remove large areas of agricultural land from production.

So we are not just talking about what some people in cool climates might see as a rather welcome process of warming. Looking just at the UK, assuming the Gulf Stream is not switched off, average summer temperatures in the UK might be higher, benefiting tourism, but against that there is the prospect of a 10 per cent increase in average annual rainfall, and possibly a 20 per cent increase in Scotland, with consequent flooding and crop damage. In addition, there is the prospect of increased storms and, at other times, droughts. The impact of climate change may not be as bad for the UK as for some other countries, but it seems it is unlikely to be welcomed. As the UK's Secretary of State for the Environment put it in 2001,

> increased risk in the UK of droughts, heavy rainfall and floods, could have major consequences for land use, planning, water resources, infrastructure, insurance, tourism and many other sectors across society. Climate change must be factored into everyday decisions by organisations and individuals now. People must not be caught on the back foot. Even at the lower end of the range of uncertainty they will have a huge impact on all our lives.

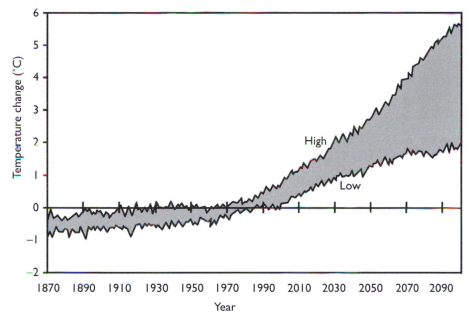

Figure 2.1 *Projections of possible global average temperature increases. Range of possible temperature changes predicted by climate models, compared with the average temperatures over the baseline years 1961–90. Based on continuing emission of greenhouse gases, but leaving out the impact of sulphate aerosols, which are thought likely to reduce the warming impact slightly (e.g. by around 0.5 °C by 2100 for the highest scenario shown here)* (*Source:* Derived from the IPCC Third Assessment Report, 2001)

The climate change debate

For some, climate change is already seen as a reality, confirmed by the increasing occurrence of storms, droughts, flooding and climate-related disasters. Of course, care has to be taken with anecdotal reports, rather than statistically significant evidence, but the last few summers in the northern hemisphere have certainly been the hottest on record, and for those facing the imminent threat of inundation by rising sea levels, there is little to debate.

However, there are still disagreements about exactly what is happening and what is likely to happen, and much general speculation. Certainly there is room for debate over the longer term prognosis, given our imperfect understanding of the very complex processes at work in climate systems, for example, in relation to the role played in climate change by clouds. The IPCC is increasingly confident that their general conclusions are correct and that the uncertainties in the predictions have been reduced, but there is a minority view that is less sure. Indeed, the so-called 'contrarian' position was initially that climate change was not in fact happening. There was much made of satellite data which showed cooling rather than warming. However, one would expect reduced radiation outwards from the planet as a result of an enhanced greenhouse effect, although this does not entirely explain the temperature anomaly in the outer atmosphere. Nevertheless, the surface data, which have been collected over a much longer period, consistently showed warming. As that became more apparent in reality, contrarians have tended to move on to claiming that there are other possible causes than human activity, one of the more prominent contrarian explanations being variations in solar activity (Calder 1997).

It is certainly true that there are periodic variations in solar activity, such as sunspots and solar flares, which can have an impact on the earth and its atmosphere, but it is hard to see how these (so far) relatively short-term cyclic phenomena could lead to the sorts of climate changes we are now facing, which seem to be unprecedented in at least the last thousand years, possibly the last 10,000 years. More specifically, the effect of changes in solar activity, combined with the dust (and in particular, sulphate aerosols) produced by volcanic eruptions over the last few decades, seems to have been to produce a cooling effect, which, arguably, has counteracted the impact of the underlying process of global warming. This brings us to another 'contrarian' view that the planet is moving towards a cooling phase (i.e. a new ice age) and that any warming that mankind is causing will ultimately be irrelevant.

A variant on that argument is that the planet is, independent of our activities, ramping up to a peak in temperature, before the next ice age. However, the rate and scale of the temperature increases that seem likely to occur, and the changes in carbon dioxide concentrations in the air that have already occurred in recent decades, are greater than those that occurred during any of the very large climate shifts that have occurred over the past 400,000 years. As Figure 2.2 clearly illustrates, the carbon dioxide concentrations have leapt up in recent years, well above the levels attained during the last four major warm interglacial periods, as revealed by the data from Antarctic ice-core measurements.

What is interesting in relation to these data, apart from the quantities involved, is that in the past four interglacial cycles, carbon dioxide levels

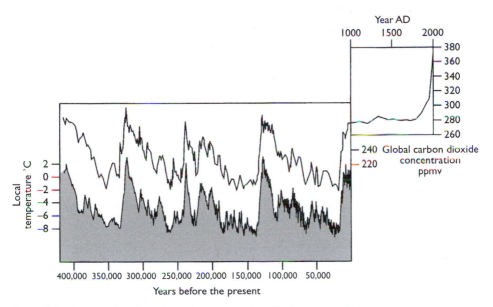

Figure 2.2 *Carbon dioxide and temperature data – the ice-core evidence*
(*Source:* Royal Commission on Environmental Pollution, 'Energy – the changing climate', 2002)

have apparently increased as a *response* to warming cycles, presumably with carbon dioxide outgassing from the seas and other carbon sinks. However, now increased carbon dioxide levels seem to be the *driver* for climate change. Isotropic measurements have indicated that there is an increasing proportion of 'old' carbon in the atmosphere, presumably from the combustion of fossil fuels.

So it seems that something new is occurring, and human causation seems a very strong contender, a view that has been strengthened by a study published in 2002 by the Goddard Institute for Space Studies, which puts the impact of other explanations, including solar variations, in perspective (Hansen *et al*. 2002).

Although the often heated debate over the causes of climate change continues, the emphasis has moved on to what might be done about it, with the costs of any response being a contentious issue. See Box 2.2. The problem facing those involved with making decisions on what, if anything, to do, is that it will probably be some decades before a full understanding of the complexities of climate systems can be reached, and by then it may be too late to respond effectively, except, where possible, by building dikes! Indeed that might not even be a viable response if some of the more extreme predictions prove to be correct. For example,

Box 2.2

The cost of climate change

It is hard to put accurate figures on the economic cost of the potential damage and dislocation likely to be caused by climate change. Just focusing on the industrial countries, the insurance company Munich Re has estimated that if carbon emission levels doubled then that would impose direct costs on the UK of around $25 billion p.a. and $150 billion p.a. for the USA, which is 1.4 per cent of the USA's current GDP. This assessment is likely to understate the full magnitude of the global problem. The former UK Government Chief Scientific Advisor, Sir Robert May, has noted that, in addition to the costs of the direct damage to property and the economy, there could be major impacts on natural ecological processes such as soil formation, water supplies, nutrient cycling, waste processing and pollination, which would have indirect economic implications around the world. He reported rough estimates of the economic value of these processes as being around £10–34 trillion per year, about twice the conventional global GNP, and noted that 'large swathes of this £10–34 trillion are at risk from the possible environmental and ecological changes sketched by the IPCC'.

Assessing the cost of trying to avoid climate change by avoiding carbon dioxide emissions is hard, since in the decades ahead all the existing energy generation systems will have to be replaced in any case, as power plants reach the end of their working lives. So that cost will have to be met come what may. Some of the new 'cleaner' more efficient technologies discussed later in this book may be more expensive, but may save money in the long term, since fuel prices seem bound to increase. However, if the changeover to non-fossil energy generation and energy- efficient end use has to be done rapidly, that could incur significant extra costs in the short term. A delayed response might therefore have its attractions politically but also, possibly, technologically. If the changeover could be left until later, cheaper and better technologies would presumably be available. Against that, the later a response is made, the worse the climate situation could be – and the greater the cumulative cost of environmental damage.

Despite that, some critics, worried about the costs of technological changes, seem prepared to accept whatever damage occurs, and want no significant response at all, other than adaption to the inevitable. Bjorn Lomborg, the author of the controversial book *The Skeptical Environmentalist* (2001) has argued that trying to limit climate change might cost $150 billion p.a., with the total cost put at perhaps $4 trillion, which he suggests is comparable with the costs that will be incurred from climate change if nothing much is done (*Guardian*, Special Report, 17 August 2001).

However, these are very speculative figures. As noted above, Munich Re has estimated that the USA alone could be faced with $150 billion p.a. damage costs, and, in practice, some of the costs of the climate control measures are likely to be offset by the economic advantages to be gained from more efficient energy usage. Moreover, if nothing is done to limit climate change, the damage costs will grow and continue, whereas the cost of changing over to cleaner energy technologies is mostly a 'one-off' cost. Certainly, as the technology improves, any extra operating costs should reduce.

On balance then, action now seems preferable to inaction, or action later, especially since, beyond some as yet poorly understood threshold levels for emissions, it seems that some aspects of climate change, once established, could take a long time to reverse and some may, in effect, be irreversible. For example, the IPCC has suggested that thermal expansion of the oceans would continue long after carbon emissions were cut and the sea levels would also continue to rise, due to the continued melting of the ice sheet, for hundreds of years. Moreover, once the ice sheets have melted, it would take several millennia for them to reform, mainly since the rate at which the seas can absorb heat is low. So, beyond a certain point, even if further carbon emissions were halted, the increased sea levels that had resulted from the melting ice-caps would be with us almost indefinitely. The window of opportunity for taking action to avoid these change may be quite small. As the IPCC note, given the inertia in climate system responses, the sooner the reduction in carbon emissions starts, the lower will be the level at which carbon dioxide levels will stabilise (IPCC 2001).

it is conceivable that when and if temperature begins to rise above a certain point, an ever more rapidly accelerating 'runaway' greenhouse effect could occur, as the warming seas are unable to absorb carbon dioxide, and more methane is produced from the spreading wetlands (Legget 1991). In addition, as the IPCC argued in their third report in 2001, it could be that, beyond certain as yet only partly understood threshold levels of carbon dioxide concentrations, some aspects of climate change could in effect be irreversible, implying a need for urgent action to reduce emissions (see Box 2.2). Certainly, to those who are convinced that human activities are the main cause, it seems likely that the sooner emissions are cut back the less risk there will be, although there are those who argue that climate change may now be unstoppable, so that we can only focus on adapting to it as best we can.

Political reactions

In the 1990s the threat of global warming and climate change was taken increasingly seriously by governments around the world, although not all were prepared to adopt the 'precautionary principle', which would suggest that action should be taken now despite the absence of full scientific identification of the scale of the problem. However, some governments seemed to be willing in principle to adopt what is called the 'policy of least regret', that is to support developments which would be sensible even if global warming did not turn out to be such a significant problem. The result is that relatively modest measures were adopted. The Rio Earth Summit in 1992 produced an agreement backed by more that 160 countries to try to get carbon dioxide emissions back to 1990 levels by the year 2000. At the UN Climate Change conference held at Kyoto in Japan in 1997, preliminary agreement was obtained for a follow-up programme, with emissions to be reduced globally by, on average, around 5.2 per cent below 1990 levels, during the period 2008–12, with each of the industrialised countries being given a target. For example, the EU's target was an 8 per cent reduction and the USA's was 7 per cent.

Unfortunately, not all of the initial signatories ratified this agreement – notably the USA, the largest single source of emissions. The USA had clearly been unhappy with the accord, and when George W. Bush was elected its opposition became more forthright. In March 2001, Bush commented

> I oppose the Kyoto Protocol because it exempts 80 per cent of the world, including major population centers such as China and India, from compliance, and would cause serious harm to the U.S. economy . . . there is a clear consensus that the Kyoto Protocol is an unfair and ineffective means of addressing global climate change concerns. . . . I do not believe that the (US) government should impose on power plants mandatory emissions reductions for carbon dioxide, which is not a 'pollutant' under the (US) Clean Air Act.

Subsequently in 2002, the USA withdrew formally. Australia also refused to ratify the treaty, although, by the time of the Earth Summit on Sustainable Development in Johannesburg in 2002, it was clear that most of the rest of the countries in the world would ratify it.

However, even if the Kyoto targets are met, this would only make a very small contribution to limiting climate change. To put the situation in context, the Intergovernmental Panel on Climate Change has suggested that net reductions in global emissions of up to 60 per cent below 1990 levels would be required in order to stabilise climate change.

More radically still, if, as many argue, the main burden of reducing emissions must fall on rich countries, which have contributed most of the carbon to date, the industrialised countries would have to cut their emissions by proportionately more, perhaps by 85–90 per cent by 2050. Clearly there is some way to go before the world community can think in those terms.

Technical alternatives to fossil fuels

There has however been some enthusiasm for a return to the idea of using nuclear power – since nuclear plants do not generate any carbon dioxide gas. As we will be discussing later, the scale of nuclear expansion that would have to occur to make a significant impact would have to be very large, and it seems unlikely that this will happen. Some countries, like France, Japan, China and Korea remain keen on nuclear expansion, and the USA has recently also begun to reconsider this option. However, in most of the rest of the world nuclear power is perceived as a failed or at least stalled option. In part the problem has been economic. The 'energy crisis' envisaged in the mid-1970s failed to materialise: fossil fuel prices subsequently fell and nuclear power could not compete. It had also become hard to support politically in many countries due to public opposition.

As Box 2.3 illustrates, the 1970s saw the beginning of objections to nuclear power from environmental groups, based primarily on safety concerns, and these were, in some people's views, confirmed by the spate of nuclear accidents – for example, at Three Mile Island in the USA and Chernobyl in the Ukraine. Popular opposition throughout the industrial world, some of it very militant, led to nuclear programmes being abandoned or slowed, and to some countries, like Denmark, deciding not to even attempt to go down the nuclear route.

Following the Three Mile Island accident in the USA in 1979, support for nuclear power in the UK dropped notably, and it fell even further after the Chernobyl accident in April 1986, with, by May 1986, only around 18 per cent of those asked in a Gallup poll supporting an increase in nuclear power, as against 39 per cent calling for 'no more at present' and 36 per cent saying 'stop nuclear power'. By 1991, a Gallup poll showed that support had fallen even further, to around 13 per cent, while 78 per cent of those asked either wanted 'no more nuclear plants at present' or for the use of nuclear power to be halted (Gallup 1994). Since then opposition seems to have reduced slightly, possibly reflecting increasing concerns about climate change, but even so, resistance remains quite strong. For example, in a poll carried out by the British Market Research Bureau in 2001, 68 per cent of those interviewed said that they 'did not think that nuclear power stations should be built in Britain in the next ten years' (BMRB 2001). A year later, in 2002, a MORI poll of UK residents found that 72 per cent of those asked favoured renewable energy rather than nuclear, while a National Opinion poll carried out in the same year

Box 2.3

Nuclear opposition

It is worth remembering just how strong opposition to nuclear power has been.
It became most visible when nuclear plant construction programmes began to expand
around the world following the 1973–4 oil crisis. In the USA there were mass
demonstrations at nuclear sites, for example, in May 1977 at the site of the proposed
reactor at Seabrook in New Hampshire, where 1,400 people were arrested. These actions
were non-violent, but in Europe larger and more militant demonstrations took place: for
example, in November 1976 30,000 people attended what was planned to be a peaceful
demonstration against the planned reactor at Brockdorf in Germany. Some 3,000 tried to
occupy the site and there were violent clashes with the police who used water cannons,
tear gas grenades and baton charges to try to restore order. Similar battles took place at
Grohnde near Hamelin in March 1977. In July 1977, during a major demonstration
against the French prototype fast breeder reactor at Malville, involving more than
60,000 people, one demonstrator was killed.

The demonstrations nevertheless continued: in September 1977, 60,000 people protested
at the site of the proposed fast breeder at Kalkar in Germany, and, in the same month in
Spain, 100,000 people joined a protest in Saragossa, while 600,000 took part in a
demonstration against plans for a reactor at Lemoniz (Elliott 1978).

Of course some of these protests might be written off as just the results of anti-
establishment agitation amongst a volatile student movement. However, opposition
to nuclear power spread well beyond the fringe. For example, in the 1980s in the UK,
when attempts were made to find sites for low-level nuclear waste repositories, they
were met by blockades by local farmers and militant opposition by residents, even in
very conservative parts of the country (Blowers 1991).

The issue of what to do with nuclear waste has continued to generate conflicts,
particularly in Germany, where shipments have been blockaded by objectors. In parallel,
transport of waste and nuclear fuel by sea has increasingly been confronted by protestors
seeking to halt this trade, who are worried in particular about the risks of seizure by
terrorists of materials that could be used for weapons.

for the Energy Saving Trust found that 76 per cent believed the
Government should invest time and money developing new ways to
reduce energy consumption, 85 per cent wanted government investment
in 'eco-friendly' renewable energy (solar, wind and water power) and only
10 per cent said the government should invest time and money in building
new nuclear plants.

Interestingly, the level of opposition in the USA, which typically has run
at around 60 per cent, initially decreased after the energy crisis in

California in 2001, when it was claimed that power blackouts could be avoided in future by building more nuclear plants. However, this argument subsequently seemed to lose its force after the terrorist attacks on the USA in September 2001. A Gallup poll showed support for nuclear power dropping from 48 per cent in May 2001 to 42 per cent in November 2001.

It is wise to be cautious about interpretations from opinion polls, not least since the results depend on the framing of the question. For example, a very different response is expected from the question 'should we build nuclear plants to stop blackouts?' compared to the response to 'should we build a nuclear plant near your home?'

There will also sometimes be different responses depending on whether the topic is existing or new plants. For example, in Sweden, where the government is trying to phase out nuclear power, a poll in 2001 found that only 19 per cent supported premature closure of the Barseback-2 nuclear plant, as planned by the government, while 37 per cent favoured continued operation of all the country's 11 nuclear power units and a further 28 per cent favoured this, plus their replacement in due course. But only 11 per cent wanted to further develop nuclear power in Sweden.

Of course in some countries there is overall support for nuclear power. For example, according to an IPSOS opinion poll reported in the *International Herald Tribune* in August 2002, almost 70 per cent of French adults have 'a good opinion' of nuclear power, although 56 per cent also said, rather fatalistically, that they believed a 'Chernobyl type accident' could happen in the country. However, although there are clearly exceptions, it does seem that, in general, with contemporary events sometimes increasing or decreasing the level of opposition or support, nuclear power is not popular with most people.

Public opposition on this scale has clearly affected decision making about nuclear power, with governments being aware of its unpopularity. More generally, environmentalist pressure on safety issues forced up the price of nuclear power, and that, along with the relatively low level of fossil fuel prices, has had the effect of further undermining the economics of the industry.

The longer term prospects for nuclear power are unclear. Fission reactors use a fuel (uranium) that, although still relatively abundant, will not be available indefinitely, and fast breeder reactors, which in effect would stretch the availability of the fuel, have yet to be operated commercially, and pose what some people feel are significant safety and security risks. Finally, nuclear fusion remains a long-term possibility, but, as we shall see later, even if the technology can be perfected, it too has its own problems.

Alternative energy options

Rather than try to create little artificial suns on the earth, in the form of fusion reactors, many environmentalists believe it is more immediately credible to make use of the natural fusion energy that the sun already produces and which reaches us as sunlight. As has been mentioned, ever since the 1973–4 energy crisis, research on solar power and other forms of renewable energy has expanded and has led to some relatively large-scale deployments. For example, by 2002 there was around 31 gigawatts of wind turbine generating capacity in place around the world.

As has been noted earlier, the term renewable energy is used to indicate that these natural energy sources (for example sunlight, winds, waves and tides) cannot be used up – unlike fossil or nuclear fuels they are not based on finite reserves, but are naturally replenished. Nevertheless, as we will be discussing in subsequent chapters, there can be problems with trying to use what are generally more diffuse and sometimes intermittent energy sources. Even so, renewable energy conversion technology is developing rapidly, and although renewables are currently only making relatively small contributions to overall world energy supplies, they look very promising in the longer term. For example, the relatively conservative World Energy Council has suggested that, given the necessary support, renewables could supply up to 30 per cent of world energy by the year 2020 and perhaps 50 per cent by the year 2100 (World Energy Council 1994). For comparison, a study carried out for Greenpeace suggested that given proper support renewables might actually supply almost 100 per cent by 2100 (Greenpeace 1993).

Energy conservation is equally promising but also still rather marginal in many countries. Energy savings of 50–70 per cent are seen as possible in many sectors through the adoption of relatively cheap measures, and yet the incentive to invest in energy saving is so far generally less than the incentive to invest in new energy supply options.

Clearly, in the case of both renewable energy and energy conservation there seems to be a need for a more forceful approach – if, that is, the global warming problem really is as significant as many believe.

There have been attempts by some governments to promote the development of renewables and conservation by the use of grant aid, subsidies and other financial incentives, and there have also been proposals by the European Commission for carbon taxes that would penalise fossil fuel use, and hence stimulate alternative energy developments. At the same time some renewable energy technologies and some energy conservation techniques are beginning to be taken up under the influence of conventional market pressures – since they are commercially attractive in some circumstances. However, not everyone is convinced that the problem of developing a sustainable energy system can be resolved just by throwing money at technology, or by imposing broad taxes, or by leaving it all to the market. What is needed, it is argued, is an overall strategy to guide energy developments.

Part 2 of this book reviews some of the technical options. But to set the scene, Chapter 3 attempts to develop some basic criteria for a sustainable energy system, as a guide for the future, drawing on the discussion so far. That may provide us with a way of assessing the various energy options.

Summary points

- The combustion of fossil fuels is creating major global environmental problems, most notably global warming.

- Alternative technologies exist which may resolve some of these problems.

- Nuclear power is seen by some as one possibility, but it has economic problems and it has met with widespread public opposition based on concerns about safety.

- Although it has its own problems, renewable energy technology, along with energy conservation, may be a better bet.

- We need to have some criteria to help us assess, and choose amongst, the various energy options.

Further reading

For an interesting history of energy use in the world over the centuries see Jean-Claude Debeir, Jean-Paul Deleage and Daniel Hemery, *In the Servitude of Power: Energy and Civilisation through the Ages* (Zed Books, London and New Jersey, 1991). And for a delightful review of the whole area of energy systems and their relations to man and the ecosystem see Vaclav Smil, *Energies: an Illustrated Guide to the Biosphere and Civilization* (MIT Press, Cambridge, MA, 1999).

The Open University Environment course reader *Energy, Resources and Environment*, J. Blunden and A. Reddish (eds), co-published by Hodder and Stoughton, London, in 1991 and revised in 1996, contains an excellent introduction to energy concepts and energy-related environmental problems. Janet Ramage's *Energy – a Guidebook* (Oxford University Press, Oxford, 1997) is one of the most useful general introductions to energy studies.

Power for a Sustainable Future: Energy Systems and Sustainability (G. Boyle *et al.* (eds), Oxford University Press, Oxford, 2003) is also a very useful source of information of existing energy systems. It forms part of the Open University course on 'Energy for a Sustainable Future' (T206).

There is a vast literature on global warming and climate change, ranging from the technical and analytical to the prescriptive. The 1992 report of the IPCC, the Intergovernmental Panel on Climate Change, contained detailed analysis of the problem and of some of the recommended policies for dealing with it; further reports emerged from the IPCC in 1995 and 2001 (the later often being labelled 'TAR', or the Third Assessment Report). See http://www.ipcc.ch.

Probably the most useful general introductory text on climate change is John Houghton's *Global Warming: the Complete Briefing* (Cambridge University Press, Cambridge, 1997). For an overview of the Kyoto accord see M. Grubb *et al.*, *The Kyoto Protocol: a Guide and Assessment* (Earthscan, London, 1999). For a critical review from a US perspective see D.G. Victor, *The Collapse of the Kyoto Protocol and the Struggle to Slow Global Warming* (Princeton University Press, Princeton, NJ, 2000).

If you want to keep up to date on the various global negotiations that are underway on emission standards and so on, you could subscribe to the Earth Negotiations Bulletin, a version of which is available at http://www.iisd.ca/linkages/.

Although some contrarian views may be sponsored by fossil fuel interests, that does not necessarily mean all divergent views should be discounted: good science progresses by informed debate and challenges to orthodoxy. For 'contrarian' web-sites and links see http://www.sepp.org and http://www.co2science.org/.

If you want to explore the past history of climate change, including the Vostok ice-core data from the Antarctic going back 400,000 years, see http://www.cdiac.ornl.gov/trends/co2/contents.htm.

To explore the potential role of changes in solar activity in relation to climate change see http://www.grida.no/climate/ipcc_tar/wg1/245.htm.

For an overview of the various factors that might be causing climate change, including changes in solar activity, see Hansen *et al.* (2002) 'Climate forcings in Goddard Institute for Space Studies SI2000 simulations', *Journal of Geophysical Research*, DOI10.1029/2001JD001143, which can be downloaded in PDF from http://www.giss.nasa.gov/gpol/papers/2002/2002_HansenSatoN.pdf.

3 Sustainable technology

- Energy resources
- End use efficiency
- Natural energy flows
- Environmental impacts
- The costs of energy
- Energy limits

All energy technologies have some environmental impacts. To help in the process of selecting which technologies might be most important for a sustainable future, this chapter develops a set of environmental criteria and strategic guidelines. They concern specific local impacts as well as overall energy resource limits. The criteria that emerge form the basis of much of the subsequent analysis in this book.

Criteria for sustainability

As we have seen, a range of environmental and resource problems have emerged so far from the use of energy technology. Our review of these issues should have indicated some basic criteria for technologies which avoid or minimise the problems.

Let us assume that our 'technical' goal is *to devise a set of energy technologies which can meet human needs on an indefinite basis without producing irreversible environmental effects.*

From this simple definition of the technical requirement, it is clear that *the technologies should not use fuel resources that are likely to be rapidly depleted.* This is the first criterion, which is concerned not with environmental impact but with fuel reserves and resource availability.

Of course, all the existing fuels (coal, oil, gas and uranium) only exist in finite amounts on this planet and are not being replaced. The conditions under which the fossil fuels were formed were probably unique to specific periods in the planet's history, and the uranium reserves are simply a mineral inheritance. That would imply that the only viable long-term energy source is renewable energy. While that may be true, the issue is the timescale that is involved here. For the immediate future, over the next few decades at least, there is no choice but to rely on fossil fuel resources for the bulk of our power – and some would say we must also continue to use nuclear fuel. But the reserves of these fuels vary. To see what the first criterion above means in practice, we need to have some idea of what fuel reserves and energy resources are available.

Reserves

Energy resource and fuel reserve issues are clearly of central importance for any sustainable energy system: we have to be able to rely on having continued access to energy sources. It might be reasonable to accept the use of fossil fuels in the interim, while alternative sources are being developed, but most of the fossil fuels do not represent long-term options.

It is hard to give precise figures for reserves. Reserves do not actually 'run out', they just become more and more expensive, until it becomes unproductive to search for and then extract what remains. In fact the main practical constraint in the short term may be the *rate* at which the reserve can be accessed, not its total scale. Although there must inevitably be a limit on how much can be obtained, the extent of economically available mineral resources depends on the economics of extraction relative to the economics of energy production and sale, and on the level of investment in exploration to find new reserves. So, the size of the economically viable global coal, oil and gas reserves depends in part on the price consumers are willing to pay, as well as on the rate of use. However, it must surely be only a matter of decades before some key reserves become expensive to access. Coal reserves should last much longer than oil and gas, of the order of two hundred years or so, but gas and oil are likely to become more expensive over the coming decades. Indeed, although new finds continue to be made and the oil and gas industries remain confident about the future, some observers believe that we have already reached the peak in extraction of oil and may do soon for gas (Campbell and Laherrere 1998).

As we shall see in Chapter 6, **nuclear fusion** is a long way from being a viable option, but it might conceivably provide power for much longer periods than fossil fuels. So might a switch over to fast breeder fission reactors, which could, in effect, stretch the world's uranium reserves. However, the global availability of the type of uranium used in conventional nuclear reactors is limited to perhaps of the order of one or two hundred years, depending on the rate of use.

Now these are relatively long timescales in everyday terms, although they are very short in terms of human history and even shorter in terms of the geological process of laying down the fossil reserves. In Figure 3.1, energy analyst Gustav Grob puts the fossil fuel era in a wider perspective. He sees it as a short 'blip', to be followed, as viable resources are depleted and/or environmental concerns rise, by a switch to renewable energy technology, using natural energy flows. Nuclear technology, whether fission or fusion, is not seen as relevant on this large timescale; for Grob, renewables are the only significant long-term, non-finite resource (Grob 1994).

However, even in the most optimistic scenario, the change over to renewables would take at least a century, so there would probably be problems with maintaining supplies of some fossil and nuclear fuels (at least, in the absence of the breeder reactor) sometime before renewables could take over fully. As we shall see in Chapter 6, even if it can be successfully developed, it seems unlikely that nuclear fusion could make a significant contribution in time to make up the shortfall. So there is a very

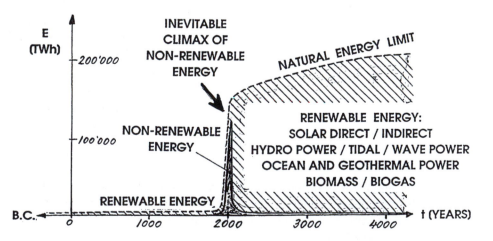

Figure 3.1 *Gustav Grob's view of the long-term patterns in global energy supply*
(*Source:* G. Grob, 'Transition to the sustainable energy age', *European Dictionary of Renewable Energy Suppliers and Services*, James and James, London, 1994)

strong case, just on resource grounds, for using the remaining fossil fuels as efficiently as possible.

So the second criterion is a strategic one: *the efficiency of energy generation and use should be improved as much as possible, as an interim measure, while the new sources are fully developed.*

Conserving energy – or generating more?

Some enthusiasts for energy conservation argue that it can achieve so much by way of energy saving at the point of use that the energy supply side becomes more or less irrelevant, at least in the short to medium term. There is talk of savings of up to 90 per cent or more in some sectors, with clever 'technical fixes' allowing us to improve the overall efficiency of energy resource use by a factor of four, as Amory Lovins and his colleagues have argued in their influential book *Factor Four*. They even suggested that factor ten savings might be possible (von Weizsacker 1994).

Certainly, if this could be achieved it would be much easier to imagine supplying the small amount left from renewable sources, or even from fossil and/or nuclear sources, while they lasted. The reality, however, is that it will take time to achieve savings of anything like this level. According to the Intergovernmental Panel on Climate Change, at best, if the currently most advanced energy-using technologies could be introduced in each 'end-use' sector, it might be technically feasible to achieve overall energy efficiency gains of 50 to 60 per cent, and, although they suggest this might be possible in many parts of the world, they see the timescale as being two to three decades (IPCC 1995). Even in developed countries progress seems likely to be relatively slow. For example, the UK Cabinet Office's Energy Review, published in 2002, suggested that the efficiency of energy use in the domestic sector might be increased by 20 per cent by 2010 and a further 20 per cent by 2020 (PIU 2002). So it could take time for emissions to be reduced in this way.

During that period, fossil fuels would still be used. Nuclear power might be able to help reduce carbon dioxide emissions during this period to some extent, but it seems unlikely that it could expand sufficiently to play a very significant role.

The result is that, unless there is also investment in renewables, all conservation does is delay the time when the fossil fuels are burnt off.

In the end all the carbon dioxide is still released into the atmosphere, albeit at a lower rate. If the rate of release of carbon dioxide could be dramatically reduced, then of course it might be possible to limit the global warming effect. However, as we have seen, by itself, energy conservation seems unlikely to be able to achieve this quickly and the overall contribution it could make may well be dwarfed by the continued rise in demand for energy. While there will hopefully be many technical and operational innovations that can improve efficiency, once the easy and cheap options have been exploited, it is hard to see how further efficiency improvements can be replicated continually, so as to keep pace with the projected 2 per cent yearly increase in basic global energy demand into the future.

Using fuels efficiently

Although there may be limits to what can be achieved by energy conservation measures on their own, improving the efficiency with which fuel is used is nevertheless vital and urgent, as part of a wider strategy for achieving a sustainable energy supply and demand system. This is particularly clear when we look at the way power generation has been carried out so far.

Historically fuels have been relatively cheap and plentiful, so overall conversion efficiency did not seem to matter too much. The giant power stations built in the post-Second World War period in the West were slightly more efficient than their predecessors, but they still only converted around 35 per cent of the energy in the fuel used into useful energy in the form of electricity. Increasing the overall size of the plant can produce marginal increases in conversion efficiency, but at best the maximum efficiency that can be obtained is around 40 per cent. This is not a technical limit but the result of a fundamental thermodynamic constraint associated with the process of using heat to raise steam to drive turbines. It applies equally to steam-raising plants heated by nuclear reactors.

Once generated there are yet further losses, of up to 10 per cent or so of the initial power sent out, in the distribution process down the national power grid. And the efficiency at the point of use is similarly low; for example, many houses are poorly insulated and domestic appliances are often very wasteful in energy terms.

The end result is that only a proportion of the energy in the original fuel is available to be employed at the point of use. The rest is often wasted heat,

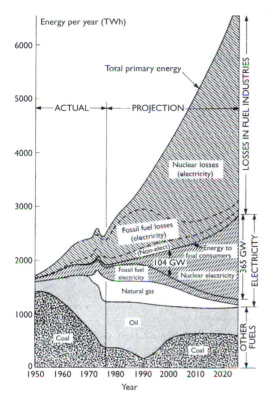

Figure 3.2 *The provision of energy in the UK, projected to 2025 on the basis of the 'official strategy' as conceived in 1976, when a major expansion in the nuclear element was being considered*
(*Source:* Royal Commission on Environmental Pollution, Sixth Report, 'Nuclear power and the environment' (The Flowers Report), Cmnd 6618, HMSO, London, September 1976, p. 180. Reproduced by permission of the Controller, HMSO)

ejected into the environment. The sheer scale of this waste is apparent from Figure 3.2, which shows the energy system at one time projected for the UK, based on a major expansion of nuclear power. As can be seen, around one-half of the input energy would be converted at the power stations into waste heat.

If this trend was continued indefinitely, ever increasing amounts of heat would be released into the environment. Microclimate effects have already been noted in the immediate locality of some power stations, and, ultimately, if this process of thermal pollution continued, serious environmental problems could emerge on a global scale, as what has been called the 'heat death' limit was approached. Note that this has nothing to do with global warming: it is just raw atmospheric heating. Nuclear fission or nuclear fusion would be no answer: they release just as much heat.

Combined heat and power

Fortunately however, use can be made of some of the heat otherwise wasted from power stations – for example, by feeding it to 'district heating' networks, with power stations operating in the so-called 'co-generation' or 'combined heat and power' (CHP) mode. Thus the heat is used, as well as the electric power output. There are many such systems already in operation around the world, particularly in Scandinavia, where many towns have district heating networks, with hot water from power stations being fed to houses and other buildings by underground pipes to provide space heating.

Obviously, given that large power stations are usually built away from centres of human habitation, there may not always be a convenient use for the large amounts of heat available – it cannot be transported great distances. However, CHP does not have to be on a large scale. Some forms of CHP involve relatively small and cheap engines, basically like car engines, while more recent micro-CHP units make use of small gas turbines, with, in both cases, natural gas being used as a fuel. The heat output from micro-turbine systems is at a higher temperature than from conventional steam-raising combustion plants, so gas turbines are well suited to CHP and, being smaller, they can be located nearer to heat demands. For example, they can provide heat and power for industrial and commercial use and for housing projects. The overall energy conversion efficiency can be up to perhaps 80 per cent or more, since most of the energy input to the power stations is converted into useful energy.

Rather than being used in this way for 'district heating' schemes, the exhaust heat from gas turbines can also be used to generate electricity. There has been rapid growth in the adoption of the two-phase 'combined cycle gas turbine' (CCGT). Natural gas is used to fuel a gas turbine and generate electricity, as with normal gas turbines, but the exhaust gases are then used to raise steam for a conventional steam turbine. So both phases generate electricity. The overall efficiency of energy conversion can be up to 55 per cent or more, much higher than for conventional coal plants or gas turbines plants, although still lower than with operation in the CHP mode. Even so, the so-called 'dash for gas' has environmental attractions, since burning gas in CCGT plants typically produces around 40 per cent less carbon dioxide per kWh generated than the equivalent coal-fired plant.

Clearly, one way or another, gas-burning technology has enormous potential – and certainly it is becoming very widely used. For example, as noted earlier, whereas 80 per cent of the UK's electricity used to come from burning coal, by 2000 it had reduced to 28 per cent, while gas provided 40 per cent. However, most of this involves CCGT plants, rather than CHP: the aim is just to produce electricity.

Matching supplies to uses

This brings us to the next issue – matching the mode of energy provision to energy requirements in society generally. It seems foolish to use high-quality energy in the form of electricity just to heat houses. Certainly, this is an expensive option, in part due to the huge losses in conversion described above, which remains true even if combined cycle gas turbines are used. Although it initially costs more to install, it is still much more efficient and, in the longer term, usually cheaper to heat your house by burning gas in a modern gas-condensing boiler in your own home, than to use electricity generated in the best modern high-efficiency, gas-fuelled CCGT plant. Gas-condensing boilers typically have energy conversion efficiencies of around 85 per cent, while CCGT plants typically only achieve around 55 per cent.

This points up our third criterion: *energy production and fuel choices should be matched to the eventual end use.*

In effect, this is a logical subset of the second criterion, as is the next criterion – concerning actual energy use by consumers. As we shall see, when we look in detail later at end use energy conservation, devices exist in most end use sectors that use a fraction of the power of currently used equipment. Increasingly energy efficiency criteria are being taken into account in the design of products and production systems – as part of the move to greener products and cleaner production processes. In parallel, buildings are being designed to minimise heat loss and the need for artificial daylighting. Taking all these types of energy saving together, the relevant criterion, our fourth, is clear enough: *energy-using technology and systems should be designed to use energy efficiently.*

Energy efficiency makes sense whatever way the power is produced, whether from fossil, nuclear or renewable sources. This is obviously true in straightforward economic terms: it is foolish to waste expensive energy. But it is also true in terms of environmental impact, and particularly in terms of local impacts. Although, as we have seen, some power plants may generate less carbon dioxide than others, and some generate none at all, there can be local impacts and health risks. The air pollution problems from coal plants are obvious enough, as are the risks of nuclear accidents, but there can also be local problems with the use of renewable sources. The energy may be free, but the land required for the technologies needed to capture diffuse natural energy flows is not. It would be foolish, for example, to cover large areas of land with wind turbines just to power inefficient refrigerators. So the efficient use of power remains vital.

Local impacts of renewables

Given that the use of renewable energy sources might be a key element in a sustainable future, it is worth exploring the issue of potential local impacts in more detail. As you will see in subsequent chapters, this turns out to be a complicated issue and may well determine the success of renewables in becoming a major energy source.

Clearly, technologies that use natural energy flows will not generate pollution or carbon dioxide, but they will have local impacts – such as visual intrusion or the disruption of local ecosystems. At the same time it has to be remembered that if the energy is generated in some other way then the global impacts are likely to be much more significant. Hence the need for some sort of trade-off.

So a fifth criterion emerges, which applies primarily to renewables but also obviously applies if other energy technologies are being used: *the local impacts associated with energy technologies should be minimised and any remaining local impacts should be traded off against the global environmental benefits of the technology.*

Making such trade-offs is difficult, not least since it is hard to compare the different technologies and different types of impact. Comparisons seem to be rather *ad hoc* and are usually site specific. To try to provide a more coherent framework, a researcher at the Open University, Alexi Clarke, has developed a methodology for examining and comparing the relative impacts of renewable sources. The overall impact depends, pro rata, on the nature and overall scale of the project, and the total amount of power produced. However, according to Clarke, in comparative terms, one of the key factors influencing the relative scale of any impact is the proportion of energy extracted from the natural energy flow.

Renewable energy flows

Energy flows in nature in a variety of ways, for example in the winds, waves and tides, and it is worth considering what happens when we make use of these natural flows. The first point to note is that the extraction of energy from natural energy flows does not significantly affect the overall thermal balance of the planet: the incoming solar energy is simply redistributed to perform other functions. In contrast to the combustion of fuels or the use of nuclear fuels, no extra energy is released into the ecosystem.

Nevertheless, the extraction of energy from natural flows may have a significant impact since some of this energy may have performed crucial functions in the local or regional ecosystem and its diversion may introduce changes elsewhere in the ecosystem. For example, with hydro schemes located in rivers, it is clear that only a proportion of the water flow can be used, otherwise there could be excessive impacts on river life downstream. As can be seen from this example, the key factors influencing impact would seem to be not just the proportion of the natural energy flow that is extracted but also the nature of the local environment. Clarke goes further and argues that the nature of the energy flow concerned is also important, and in particular the energy density (or more strictly the energy flux density) of the flow (Clarke 1994).

This point can be illustrated by looking at a range of technologies and their impacts. Solar collector devices can be used to absorb heat from sunlight falling on them, but they only absorb very small amounts of the diffuse, low-energy density, solar energy flow, the result being that the environmental impact is low. By contrast, high-head hydroelectric dams attempt to intercept a large proportion of large, high-density energy flows and have correspondingly large impacts on the natural environment. Most of the other renewable energy technologies that are based on natural flows (e.g. wind, wave and tidal power technology) are strung out somewhere in between these two extremes.

Local environmental limits

The nature of the natural energy flow will need to be taken into account in the process of selecting, designing and locating systems for extracting this energy. In particular, there is a need to relate the way in which specific technologies interact with the natural energy flows in their local contexts (Clarke 1995).

Not all environments will have the same sensitivity to the presence of technologies that extract energy from local energy flows. In some cases the local ecosystem may be able to cope quite well with even high levels of energy abstraction; in other cases it may be that even low levels of abstraction are disruptive.

To complicate matters further, some parts of the ecosystem may be more sensitive than others. For example, in some cases specific animal species are likely to be more sensitive to environmental changes than inanimate geological features, although morphological changes can themselves have an impact on animal and plant life, and in some cases it will be the inanimate part of the environment that is most sensitive to change.

To explore these issues and assess the significance of any impacts, there is evidently a need to distinguish between the various elements in specific local environments – most obviously between the animate and inanimate parts, for example, animal and plant life and geological structures. These categories themselves can also obviously be subdivided, not least to separate out human beings, who have possibly unique perceptual interpretations of the significance of impacts on the environment, as well as having more direct economic or aesthetic interests in the significance of any changes. You might argue that human interpretations of what represents a good environment are often somewhat biased, for example, towards human concerns such as agriculture, access for leisure pursuits, or for the experience of natural beauty.

In order to progress further with this type of analysis, we would have to begin to consider the conflicts and possible trade-offs between human and environmental interests discussed in the model outlined in Chapter 1. But for our immediate purposes, the analysis can be simplified and we can derive a generalised sixth criterion, as a subsidiary of the fifth: *do not extract more energy from natural flows than the local ecosystem can cope with*.

Land use

To be comprehensive in our survey of environmental impacts we also need to explore the issue of land use. Land use is perhaps the most concrete way to assess the merits of energy systems on a quantitative basis. After all, it has the most visable impact on the environment. Given that there are competing uses for land, including obviously food production, but also increasingly housing, industry and leisure, this criterion could become more important. Since most renewable energy flows are diffuse, the collection technologies are likely to take up more room than conventional energy technologies. While the energy sources like wind may be renewable, land is a valuable and limited resource.

Of course, with conventional fossil and nuclear power systems we have to take into account the land areas used for mining the fuel, processing it and dealing with any wastes. On this basis, the differences between conventional and some renewable systems may not be so great. However, it is clear that, for example, energy crops will take up a lot of room.

Of the renewables, energy crops, such as short rotation coppicing of willow, are in fact the most land-using option. Depending on location, coppices can take up to twenty times more land per kWh eventually generated than wind farms, and this may matter if land is scarce. Although they are currently still expensive, electricity-producing photovoltaic solar cells (PV cells) are also a better option than energy crops in land-use terms. That is not surprising when the energy conversion efficiency of PV cells (up to 15 per cent) is compared with the efficiency of the photosynthesis of light-to-energy in biomass (less than 1 per cent). Moreover, in practice PV cell arrays can be mounted on roof tops, so there is actually no real land-use implication. According to a report by Hydro Quebec, PV, at up to 45 km^2/TWh, comes out better than wind, at up to 72 km^2/TWh and both are much better than energy plantations, at up to 533 km^2/TWh (Hydro Quebec 2000). Although energy crops can be stored, which gives them an advantage over intermittent sources like wind and solar, they are a bulky fuel, which usually has to be transported to combustion plants, and this can have implications for local road infrastructures.

Hydro dams have obvious land-use implications (since areas are flooded to make reservoirs), a problem not shared by tidal barrages. However, the latter can have significant impacts on the local and even regional ecosystem, so some land-use changes might occur, although some impacts may actually be positive. Offshore wind, wave and tidal current devices have no land-use implications, although, if they are near shore, there can be visual intrusion issues.

While the visual impacts of the offshore options are likely to be very low, wind farms on land are seen by some as ugly and intrusive, and, as we will be seeing in later chapters, that has led to local opposition to some projects. Energy crop plantations may also prove to be unpopular if they cover large areas. However, these are human perceptual and aesthetic judgements, possibly also reflecting instrumental concerns (e.g. the belief that house prices will fall).

Here we are looking at subjective, perceptual concerns, and to debates over values and, in the end, personal and political priorities. It could be argued that if people want more energy they must be prepared to accept some form of intrusion and that, for example, wind farms are one of the most environmentally benign options (as well as being economically attractive). This argument may, however, not be accepted by those who see any disruption to treasured views as deplorable, and 'anywhere but here' as a viable policy for sustainable energy. Nevertheless, there is clearly a problem of aesthetics to address and a need for careful location and sensitive consultation – something we will be returning to later. For the moment though, we will simply add a further criterion, number seven, requiring the careful assessment of local attitudes and views as part of the decision-making process concerning energy choices: *take account of local views over land use and the impacts of amenity loss.*

Overall energy limits

The next criterion for sustainability is much wider and even more complex, and may conflict with some of the others: it concerns overall energy resource limits. Looking back at Grob's chart (Figure 3.1), you will see that he has indicated a 'natural energy limit' for renewables, by which he means the maximum level of energy provision that renewables can supply.

Grob's natural energy limit concept needs some further analysis to see the full implications. Basically, it implies an absolute limit to how much energy can be extracted from natural energy flows and processes. Now there are more or less fixed amounts of overall average energy flowing into and around the planet, but the amount that can usefully be extracted will depend on the technology. Currently, most extraction techniques have relatively low efficiencies – 10–30 per cent or so – but they might improve. That sets an overall technical limit to how much power can be obtained. There are also, as we have seen, environmental limits; eco-systems can be damaged if too much energy is extracted from natural flows. Even so, the use of solar, wind, wave, tidal and biomass (i.e. biological sources of energy) will still provide plenty of room for growth – up to and beyond 200,000 TWh p.a. according to Grob, compared to the 100,000 TWh p.a. currently consumed globally. But beyond that there may be a limit to energy availability. Estimates vary as to what the limit actually is – some put it much lower than Grob, others much higher – maybe ten times more, or even twenty. But the simple point is that there are limits.

Of course, energy may not be the determining factor limiting human activity on this planet. That also depends on the complex pattern of other interactions with the rest of the ecosystem. So far the ecosystem limits are not fully understood. A whole set of ecosystem-related criteria may have to be developed – for example in terms of biodiversity and the ability of the planet's ecosystem to cope with human activities – its so-called carrying capacity.

Even so, although energy availability is not the only factor involved, the natural energy limit concept does at least provide some sort of idea of the ultimate constraints on human activity on this planet and some feel for the role of technology in responding to environmental constraints.

It leads to the eighth criterion, which concerns not technology so much as the nature of human activity on this planet: technologies *should be devised which ensure that human activities stay within the energy limits and carrying capacity of the planet*. That implies that there is a need to decide what are its limits, an issue which will be discussed in more detail later.

The costs of energy

So far we have avoided what is the most obvious criteria used at present for choosing amongst energy technologies – the cost of the various options. The economic cost of energy as charged to consumers rarely reflects the full environmental cost imposed by the process of energy conversion, but it does offer some sort of measure of the overall resource efficiency of the conversion process. Economic assessments, in terms of the cost of energy produced, are also the most obvious way to try to compare the viability of new energy systems.

However, making judgements as to which options to emphasise based on prototypes at the early stages of development can have its limitations, as we will see in later chapters when we look at the early days of renewable energy. There is also the famous prognosis that nuclear power would be 'too cheap to meter'.

Nevertheless, it is possible to identify trends. The Energy Review carried out by the UK Cabinet Office's Performance and Innovation Unit (PIU) in 2001 made use of 'learning curve' analysis to try to identify trends in price reductions. If prices at successive stages in the innovation process are plotted against cumulative production volume on a log–log scale (i.e. with the data in both axes presented in logarithmic terms), then in many cases a straight line results. The slope varies with the technology, but the range is not great – gradients of between 10 and 20 per cent are typical. Not all the energy options reviewed by the PIU could be assessed in this way, since, for the newer options like wave and tidal current technology, the data were not yet available, so parametric engineering assessments and proxy assessments had to be used (PIU 2002).

Interestingly, the PIU found that, despite very large-scale funding over a long period, the nuclear learning curve slope was only 5.8 per cent, which they attributed to the fact that it was a 'mature' technology involving large, inflexible projects with long lead times. They also argued that the frequent emergence of completely new designs meant that there was less technological continuity, less opportunity for economies of production scale, and less opportunity for learning. It does seem that nuclear power is an exception to the norm. By contrast, the results for some of the key new renewables were much more encouraging: the learning curve slopes for PV solar and wind were put at between 18 and 20 per cent.

The final conclusions for the energy options reviewed by the PIU are shown in Table 3.1. Clearly, these long-term estimates are speculative and rely on a range of assumptions about policy developments. For example, if funding is not provided for new renewables or new nuclear technologies, then the picture could look very different. But the simple message from Table 3.1 is that, although some of the renewable energy options may be relatively expensive at present, they are likely to get cheaper, as will CHP. By 2020, most will be cheaper than coal and nuclear, and some will be cheaper than gas.

Table 3.1 *Cost of electricity in the UK in 2020 (pence/kWh)*

On-land wind	1.5–2.5
Offshore wind	2–3
Energy crops	2.5–4
Wave and tidal power	3–6
PV solar	10–16
Gas CCGT	2–2.3
Large CHP/co-generation	under 2
Micro CHP	2.3–3.5
Coal (IGCC)	3–3.5
Fossil generation with sequestration	3–4.5
Nuclear	3–4

Source: UK Government Cabinet Office, Performance and Innovation Unit, Energy Review, 2002

Environmental costs

One of the main uncertainties in terms of longer term costs is over how the relative environmental costs of the various options will be assessed in future. If, for example, the full social and environmental cost of carbon dioxide and acid emissions from coal combustion is added to the cost of generation, then the comparisons would alter dramatically. There are also environmental costs with the other energy options. For example, although nuclear plants do not generate carbon dioxide there are other impacts and risks with nuclear power. Moreover, even the most benign renewable sources will have local impacts.

In effect what we are looking for is some way to operationalise the criteria we have developed in this section – turning the broad environmental assessments and the choices they imply into quantifiable costs. This is not an easy task, given the wide range of environmental impacts and the difficulty in assigning economic values to them.

A recent attempt was made in the EU EXTERNE study, which tried to put a price on the environmental and social impacts of energy use, focusing on electricity production in the EU. The results suggest that the cost of producing electricity from coal would double and the cost of electricity production from gas would increase by 30 per cent if external costs such as damage to the environment and to health were taken into account. It is estimated that these costs amount to around 1–2 per cent of the EU's Gross Domestic Product (GDP). Note that this does not include the cost of damage caused by global warming, which is not surprising, given the difficulty of coming up with estimates for this, as discussed in Box 2.2.

The methodology used to calculate the external cost is called impact pathway methodology. This methodology sets out by measuring emissions (including applying uniform measuring methods to allow comparison), then the dispersion of pollutants in the environment and the subsequent increase in ambient concentrations is monitored. After that, impact on issues such as crop yield or health is evaluated. The methodology finishes with an assessment of the resulting cost.

The EXTERNE methodology could be applied to other energy-related sectors like transport. In fact, preliminary work has shown that aggregated costs of road transport, the dominant source of damage, add another 1–2 per cent of GDP to the bill. The results for electricity are shown in Table 3.2.

As can be seen, on the basis of this analysis, fossil fuels have much larger environmental impacts than any of the other options, with coal clearly being the worst. By contrast, wind is relatively benign, having four times lower environmental costs than nuclear.

Table 3.2 Extra cost resulting from environmental damage (euro/kWh)

	Cost to be added to conventional electricity cost*
Coal	0.057
Gas	0.016
Biomass	0.016
PV solar	0.006
Hydro	0.004
Nuclear	0.004
Wind	0.001

* Assumed as 0.04 euro/kWh average across the EU (approx 2.5p/kWh)
Source: EU EXTERNE study (EXTERNE 2001)

Assessments like this are fraught with methodological problems, such as how to calculate the cost of specific types of damage. Simple economic assessment, based on insurance replacement costs, may not provide a realistic measure of, or proxies for, the human value of amenity loss or health damage, much less the ecological value of any disruption. Even more contentious is the value put on human life, which, inequitably but seemingly inexorably, differs around the world (Krewitt 2002).

Clearly then, some of these valuations are subjective and this is a problem which has perhaps even more influence on popular assessments of the local impacts of renewable energy technologies. However, having some sort of estimate for the environmental costs, to put alongside the purely economic cost, can be helpful in making choices amongst the various technologies: so our last but one criterion is *attempt to factor in environmental costs for the various options, to set alongside the purely economic costs*.

Carbon accounting and energy analysis

Rather than trying to assign costs to impacts, another increasingly popular approach to attempting to reflect the environmental significance of energy technology is to use the resultant carbon emissions. If nothing else, this might help reduce the level of subjectivity involved in making comparisons between the various energy options. In effect we are thus moving one step backwards in the EXTERNE analysis.

Certainly carbon emissions are a central factor in climate change, and it could be argued they can be used as a proxy for most other types of impact. One early estimate (Meridian 1998) put the total fuel cycle emissions for coal-fired plant at 1,058 tons of carbon dioxide per gigawatt hour, and 824 tons for combined cycle gas-fired plants, compared with nuclear at 8.6, wind at 7.4 tons and hydro at 6.6 tons. A more recent life-cycle study by Hydro Quebec, published in 2000, covered all greenhouse gas emissions and translates them into equivalent carbon dioxide terms. It put emissions at 974 tons/GWh for coal plants and 511 tons/GWh for combined cycle gas plants, compared with 15 tons/GWh for both nuclear plants and hydro dams and 9 tons/GWh for wind plants (Hydro Quebec 2000).

A study of greenhouse gas emissions by the International Atomic Energy Agency in 2000 summarises the ranges as follows: gas plant 439–688 grams of carbon dioxide equivalent/kWh, coal plants 966–1306 g/kWh, nuclear plants 9–21g/kWh and wind plants 10–49 g/kWh (IAEA 2000).

To be comprehensive, carbon accounting must include not just the emissions during operation, but also the emissions associated with the energy use in construction of the plant and the materials used in its construction, as well as the energy used in decommissioning and (where relevant) waste disposal. Full life-cycle energy analysis of this sort is becoming increasingly important for all products and systems as part of their environmental assessment. In the case of power plants, use is often made of the energy output to input ratio, that is the energy produced during the lifetime of the plant, divided by the energy required to build it – so the larger the number, the better the plant. The results can be revealing. For example, the Hydro Quebec study mentioned earlier suggested that the overall energy output to input ratio for nuclear was only 16, compared to 39 for wind, that is, over their useful lifetimes, nuclear plants generate 16 times more energy than is needed to build them, while wind turbines generate 39 times more energy than is needed for their construction. The energy output to input ratio for coal plants was 11, gas/CCGT 14, while that for hydro dams was put at 205, presumably because of the long lifetimes and large outputs of the plants once built. PV solar and biomass plantations had the worst ratios at 9 and 5, respectively, reflecting the large energy debt incurred in PV cell manufacture and the mechanical energy used with the harvesting and transportation of energy crops (Hydro Quebec 2000).

Although the energy sources may be renewable and carbon free, as can be seen, currently, some renewables, notably PV and energy crops, require more energy inputs per unit output than fossil and nuclear sources, and this may be significant if the input energy comes from the use of fossil or nuclear fuels.

Carbon accounting is becoming increasingly popular given the various proposals for carbon emission permits and carbon trading arrangements that emerged from the Kyoto climate change accord. However, while comparisons of the carbon and greenhouse gas emissions are useful, and central in terms of global climate change, there are also clearly other impacts to consider in order to give a complete picture, not least acid emissions and radioactive emissions. There is thus a danger that seeking to optimise for low carbon emissions may in fact be sub-optimal for the environment as a whole. Of course, it could reasonably be argued that climate change is so important that other issues must take second place.

However, in terms of assessing the various low carbon options, impacts other than carbon become significant. This is clearly the case in terms of radioactive pollution from the nuclear fuel cycle. In addition, since most renewables have low carbon emissions, local ecosystem effects (e.g. in relation to disruption of wildlife, biodiversity and, in the case of water flows, erosion) could be more important in choosing amongst them, which brings us back to the analysis of the impacts of abstracting energy from natural flows earlier in this chapter.

To summarise, even though carbon emissions are clearly important, we also need to keep the other impacts in mind. So our final criterion is that we should *assess the carbon emissions associated with each energy option, while remembering that other emissions may be equally important.*

Choices for a sustainable energy future

All technologies have impacts, and that implies that there is a need to make choices. In this chapter we have seen that energy technologies can be ranked and assessed in a variety of ways, most obviously in terms of their basic costs, but also on the basis of their environmental costs. However, we have seen that it is not easy to come up with hard numbers for these costs. Some people see carbon dioxide emissions as not only the crucial environmental issue, but also as a useful proxy for other environmental impacts. Nuclear and renewables come out well in this comparison. However, this type of comparison leaves out the potential for major nuclear accidents and associated health risks. By contrast, it is hard to see how many of the renewables, large hydro apart, could impose risks on the general public on the same scale as do nuclear power plants, a topic we will be returning to later.

While recognising the shortcomings, we have in this chapter developed a series of assessment criteria which attempt to alert us to most of the key environmental constraints on energy generation and use. Box 3.1 draws all of these elements together into a checklist.

If acted on fully, the ten criteria summarised in Box 3.1 could lead to a shift to a steady-state energy system with supply matched to end use demand and impact minimised to be within levels acceptable by the local and global ecosystem. Obviously it is a highly idealised set of criteria: in reality the trade-offs would be difficult and the acceptable threshold levels

Box 3.1

Criteria for sustainable energy technology

In choosing, developing and deploying new energy technologies there is a need to:

1 avoid the use of fuels which will run out;
2 improve the efficiency of energy generation and utilisation as much as possible, as an interim measure, while the new sources are developed;
3 match energy production and fuel choices to the eventual end use;
4 design energy-using technology and systems to use energy efficiently;
5 minimise the local environmental impacts of energy technologies and trade off any remaining local impacts against global environmental benefits of the technology;
6 avoid extracting more energy from natural flows than the local ecosystem can cope with;
7 take account of local views over land use and the impacts of amenity loss;
8 devise technologies so that human activities stay within the energy limits and carrying capacity of the planet;
9 attempt to factor in environmental costs for the various options, to set alongside the purely economic costs;
10 assess the carbon emissions associated with each energy option, while remembering that other emissions may be equally important.

for impacts are unknown in many cases. Longer term implications are also difficult to predict. Sustainable development is usually defined in terms of ensuring that future generations are not disadvantaged by current activities, but in practice this is very hard to foresee in every case. It is fairly clear that releasing the carbon trapped in fossil fuels is not going to be very helpful, and that bequeathing future generations with nuclear waste to deal with is not going to be well received. But what about tidal barrages for example? A tidal barrage on, say, the Severn estuary could generate 6 per cent of UK electricity and within a few decades the capital costs would be paid off. So future generations would have the benefit of a cheap renewable source, much like hydro. However that might not be welcome if it has local ecosystem effects.

Tactical and strategic issues like this abound in the sustainable energy field. In Part 2, armed with the criteria, we turn to look at the specific technical options to see how they match up to the criteria, before looking at some of the tactical and strategic issues in detail in Parts 3 and 4.

Summary points

- There are limits to fossil and nuclear fuel reserves, as well as impacts with using them.

- Consequently, when choosing energy technologies there is a need to consider the efficiency of energy conversion and use, and to use renewable sources wherever possible.

- If renewables are to play a significant role, then the local impacts of renewable energy technologies must be weighed against their global environmental benefits with land use being a key issue in some contexts.

- The world's total natural renewable energy resource has limits which may ultimately limit the level of global economic growth that is possible.

- The criteria developed in this chapter, including economic criteria, should help in the assessment of the merits of the main sustainable energy options.

Further reading

The fields of environmental economics and environmental impact assessment are expanding, and there is no shortage of texts and journals. One of the classic studies was Olaf Hoymer's *The Social Costs of Energy Consumption* (Springer, London, 1988). The 1992 report by David Pearce and colleagues for the UK Department of Trade and Industry *The Social Costs of Fuel Cycles* is also useful. The more recent series of EXTERNE reports produced by the European Union brings the analysis up to date. However, the approach adopted in this chapter cuts across some conventional analyses. If you want to follow this up, in addition to the 'reader' from the Environment course mentioned at the end of the previous chapter, the Open University has published a number of specialist reports produced by the OU Technology Policy Group analysing the interaction between natural energy flows and human interventions. See, for example, Alexi Clarke's 'Comparing the impacts of renewables: a preliminary analysis', TPG Occasional Paper, 23 December 1993; 'Environmental impacts of renewable energy: a literature review', TPG Report, May 1995.

Part 2 Sustainable technology

Part 2 looks at some of the key technologies
that might be candidates for playing a role in
an environmentally sustainable energy future.
It looks at the various energy-saving options,
at the development of 'green' consumer
products and domestic energy conservation
techniques, and at some of the ways in which
fossil fuels might be used more efficiently.
It also looks at the other energy supply
options – nuclear power and renewable
energy technology.

4 Greening technology

- **Technical fixes**
- **Clean technology**
- **Energy fixes**
- **Reducing demand**
- **Sustainable energy**

Some environmental problems can be resolved by relatively straightforward technical changes and adjustments – by what are sometimes called **technical fixes**. This chapter looks at some examples of attempts to make 'green products' available to domestic consumers, with energy saving being one of the aims. It also looks at technical fixes in the energy generation field and in transport. But as this chapter argues, while some technical fixes may be helpful as a way of reducing energy use and pollution, on its own the technical fix approach may not be sufficient to ensure environmental sustainability. More radical approaches may be needed, especially in relation to energy generation.

Technical fixes

As has been indicated, energy provisions are central to industrial and economic activity – and they are probably the key technologies in terms of environmental impact. The use of energy is determined by activities in sectors other than energy generation, for example, transport, farming, housing, production and so on. Many of the problems with the technologies in these sectors are related to their direct energy use (e.g. emissions from cars), or their indirect use of energy (e.g. pollution from the extraction and processing of raw material used for manufacturing consumer appliances).

Remedial technological solutions exist in most sectors, some of them being basically technological modifications or additions designed as a response to the problems created by the existing forms of energy generation and energy-using technology. Some of these solutions might be thought of as **technical fixes** in that they may address the symptoms, but often do not deal with the underlying causes. Thus some technical fixes attempt to clean up the emissions from power plants or cars, but do not change the basic technology which produces the emissions. The result can be that while the fix may work for a while, the problem can re-emerge in another form. As you may recall from Chapter 2, that happened when the 'fix' of tall chimneys, introduced to disperse acid gases away from power plants in the UK, subsequently resulted in acid rain damage in Scandinavia.

In the energy supply sector, remedial technical fix approaches cover each phase of the energy production process, including improvements in the handling of wastes and spoil from mining operations, the development of safer oil tankers, improvements in the efficiency of power plants and pollution control devices to filter emissions. Given that, even in the most radical alternative energy scenarios, fossil fuels will continue to play a major role in the world energy system for some while, improvements in extraction techniques, plant safety, fuel handling, emission reduction and waste disposal will be of major importance as a way to minimise environmental risks. There will also be a need to develop acceptable ways of storing nuclear wastes. These are some of the key problems facing engineers at present. Indeed, this whole book, and more, could be devoted to exploring the remedial techniques for dealing with these problems. In particular, with coal still being widely used and its use actually expanding in some developing countries, and with gas and oil likely to remain the dominant fuels for many decades, the development and application of pollution control techniques and systems will inevitably be a major focus for many technologists.

However, there are plenty of texts on such issues (see the Further Reading section at the end of this chapter), and the focus of this book is wider and longer term. For while finding technical solutions to these problems is obviously important, there is also a need to go beyond ameliorative technical fix measures designed to compensate for, but not necessarily eliminate problems, and develop more radical technical approaches that might avoid some of these problems. With this in mind, rather than exploring the engineering aspects of conventional energy technologies, this book focuses on the development of new, more radical, approaches to energy generation and use. That still of course means we will be looking at a range of technical fixes. Indeed, we will be devoting much of the rest of this book to looking at what they can achieve. However, our emphasis, especially in later chapters, is on more radical, longer term approaches – what you might call **sustainable fixes**.

Green consumer products

To start off our exploration of technical fixes, let us first look at some of the developments that have been underway in relation to consumer products. Perhaps the most familiar current examples of technical fixes are the various new environmentally friendly domestic consumer products, many of which are designed to use energy more efficiently: washing machines, refrigerators, TVs and so on. In parallel new production systems are being developed which are more energy efficient and less polluting. The term 'green product design' is sometimes used to describe the former activities, while the latter is sometimes referred to as being part of a move towards developing 'clean technology', meaning 'clean manufacturing process technology'.

Energy saving is only part of the aim: the goal of green product design and clean technology development is wider, involving the reduction of pollution, toxic emissions, and environmental impacts generally. Of course this can often only be partially achieved: rather than using the terms 'green' and 'clean' in reality what we really should be saying is 'greener' and 'cleaner' technology.

As part of this process of developing cleaner and greener technology, there have also been attempts to redesign products to use less materials or to substitute new materials for traditional ones. In part, this has been done to reduce the energy used in the manufacture of materials. For example, energy is needed to manufacture the materials used to construct cars, so that cars are said to 'embody' a significant amount of energy beyond what they use directly as a fuel. Generating the energy embodied in a traditionally designed car creates around 10–15 per cent of the carbon dioxide emissions that will be produced by burning petrol in the car during its lifetime of operation. To reduce this problem, new less energy-intensive materials are used and cars are being designed so that the material can be recycled. Many other consumer products have also been redesigned for ease of recycling, so as to be able to get access to the materials they contain, for subsequent reuse. Here again energy is a key factor: it usually takes much more energy to create new materials from virgin ores than to reuse existing materials.

In general, there is now much more attention being paid to the overall product life cycles, for example, by making products easier to renovate or repair, rather than dispose of and replace, with energy and material conservation being key factors. It is an important part of life-cycle analysis, the process of assessing the cradle-to-grave environmental impacts associated with the production, use and disposal of products (McKenzie 1991).

However, as the example of the car above indicated, what life-cycle analysis usually indicates is that, although, for most consumer items, the energy embedded in the materials in the product can be large, this is a one-off energy debt, and the main source of environmental impact is still the much larger amount of energy that they consume over their lifetime of use. Consequently, one of the main aims of 'green product designers' is to reduce direct energy consumption.

Energy-efficient consumption

As has been indicated, there is a range of technical fixes that focus on reducing energy use by consumers – by offering them new, more efficient, appliances. By purchasing such devices, consumers can have a direct influence on an important element of energy use. However, there may be limits to how much can be achieved by consumers, even if they are willing to change their consumption patterns.

A study of the purchasing and lifestyle choices made by a typical British family indicated that they could only reduce the amount of carbon dioxide emitted as a result of their activities by around a third. They could for example choose more efficient domestic heating equipment, buy a more energy-efficient car, and use public transport more. But most of the rest of the emissions came from activities over which they had no direct control as consumers: for example, the operation of power plants by the companies that supplied them with electricity, and the energy use patterns adopted by industry and commerce. The end result was that even if the family could cut their direct contribution of carbon dioxide production by a third, the net impact on total energy consumption related to their activities as consumers was only a 10 per cent reduction (Schoon 1989).

This study was carried out in 1989. Since then, new opportunities for consumer choice and influence have emerged. For example, consumers can choose to contract to green power schemes, so that the power they use is matched by energy supply companies by power bought in from renewable sources. So far the level of take up by consumers has been relatively low (60,000 in the UK by 2002), but consumer demand for green power does offer one way in which consumers can influence the decisions taken by power companies about how to generate power. We will be looking at this idea in more detail later.

Some progress has also been made in terms of recycling consumer waste, although the UK still lags far behind most other industrial countries. The environmental and energy benefits seem to be clear: it often requires more energy to produce new materials like metal and glass than to reuse existing materials; energy can be saved by recycling materials from domestic wastes; and consumers can play a role, via bottle and can banks and so on. However, the energy savings can be overestimated. The energy used in driving to recycling centres has to be taken into account, unless you combine this with your normal shopping trip.

Some of the other measures being considered as an aid to energy and material saving in the consumer product sector also need careful assessment. For example, it seems reasonable to press for long-life products rather than products that are thrown away regularly. And yet in some cases consumers might be able to buy new improved products which may be much more energy efficient: given rapid technological advance, in some contexts rapid product replacement may actually be a better way to save energy.

So, although there may be benefits, in some cases there can be contradictions: what might seem to be an environmentally sound step, might in the longer term prove to be counterproductive, or require further technical fixes. For example, the replacement of CFCs in refrigerators by gases that are 'ozone friendly' may help to deal with the ozone depletion problem in the upper atmosphere, but, unless other measures are adopted, the overall energy efficiency of the refrigerator may be lessened, and supplying the extra energy needed will lead to increased carbon dioxide emissions (Houlden 2000).

This sort of contradiction is, of course, what you would expect from a technical fix, at least in the sense that this phrase is used in this book. A technical fix implies a technical solution to a social or environmental problem which, while it may solve the specific problem, may create others, or simply push the problem on elsewhere. A more comprehensive solution may involve a more fundamental technological change or, more likely, will also involve changes in the way people use technology, in their expectations of it and therefore in consumption patterns. Indeed, it may also require wider changes in economic and social patterns.

To develop our discussion of technical fixes, let us look in a little more detail at another example drawn from the consumer product area, namely the motor car, which is, after all, one the most significant purchasing decisions, in terms of environmental impact, that consumers are likely to make.

The green car

In recent years there has been a growing concern amongst environmentalists and some cars owners about the pollutants that cars emit into the atmosphere. Governments have responded with legislation over air quality and with stricter emission standards. This, in turn, has led manufacturers to develop a number of technical fixes.

One of the first was the development of lead-free petrol – since lead emissions were seen as damaging children's health. The problem is that unleaded petrol is slightly less efficient as a fuel, so typically cars use more fuel and generate more carbon dioxide. Catalytic converters can have a similar effect: they extract some harmful gases but may reduce engine efficiency and certainly they do not reduce carbon dioxide emissions (Nieuwenhurst *et al.* 1992).

An obvious solution is to opt for electric cars. But at present these would have to use batteries recharged with electricity generated by conventional power stations. The emissions of carbon dioxide and pollutants at street level might be reduced, but some of these emissions would simply emerge at the power station. Given the fact that electricity production in conventional power stations is very inefficient, a system based on electric cars could actually generate more carbon dioxide net than using fuel directly in the car.

The limits of technical fixes can be seen clearly in these examples. So can some possible solutions; for example the use of renewably generated electricity for electric cars. However, even that is a form of technical fix: the pollution problem may have been resolved, but the result could be queues of electric vehicles. The congestion problem is not addressed, nor the damage done by building roads. To resolve these problems would need much more than technical fixes for cars: it would need overall changes in the transport system (e.g. more public transport, more use of trains and buses, more walking and cycling). That might also imply changes in living and working patterns: more use of telecommunications rather than transport; new spatial patterns of residential, commercial and industrial location to make access to work and play less reliant on the car.

Transport policy

The transport issue is a major one, and to address it fully would take us well outside the remit of this book, since it touches on so many aspects of the environment and public policy. Even if we just stay with the energy-related issues, it is clear that the transport question is becoming increasingly significant, and that it will become even more so. Energy use by private vehicles is growing rapidly around the world and it is contributing to major air-pollution problems, as well as to climate change. There are increasing calls for investment in public transport, buses and trams and trains, rather than in cars.

Air transport also represent a problem: air traffic is responsible for around 10 per cent of the total global-warming effect (including around 3 per cent of carbon emissions) and its impact is increasing. Technical improvements can certainly help to increase fuel efficiency and that should limit emissions. However, these gains can be undermined by changed patterns of use. For example, although the world aircraft fleet has doubled its fuel efficiency over the past 30 years, global air traffic has quadrupled since 1970, from 350 billion passenger miles to 1,500 billion passenger miles a year, and is forecast to more than double or even triple by 2050. To put the issue starkly, while typically fuel efficiency has risen by around 3 per cent p.a., in 2000 the use of aviation fuel in the UK rose by 10 per cent. According to the IPCC, carbon emissions from this sector could increase by 478 per cent between 1992 and 2050. Unless demand is somehow curbed, it would require major technical fixes and efficiency improvements to reduce this.

There have been calls for aircraft fuel to be taxed – at present it is one of the few commodities that has escaped any sort of tax. Cheap air flights have clearly been one reason for the boom in air traffic and consequent emissions. However, before adopting a completely anti-plane approach, it is worth noting that, perhaps surprisingly, modern aircraft, like the Boeing

777, consume only about as much energy per passenger mile on average as modern high-speed trains, so for some journey lengths there may be not much to choose between these two options, at least in terms of fuel used and the resultant emissions (Olivier 1996).

In the final analysis, however, what matters in terms of energy use and consequent emissions is not the technical route adopted, so much as the level of demand from consumers. As we have indicated, some technical fixes exist for some of the transport problems, but these may be undermined by consumer behaviour. That was what we have seen in the case of air traffic. It also is the case in terms of cars. The fuel efficiency improvements that have resulted from the relatively high level of fuel tax in Europe (compared with the USA), and the tight emission standards in the USA, have been undermined to varying degrees by consumers' insistence on switching to inefficient recreational vehicles.

Clearly, resolving the transport dilemma is not going to be easy, given the extent to which existing patterns of transport are embedded in modern society. It has been said that the advent of the car has changed society more than most political changes: so if the car is now seen as a problem, then there is probably a need for significant social changes, as well as technical changes, in order to redeem the situation (Peake 1994).

Given the central role that cars play in the economy of the developed countries, perhaps here is where the conflicts outlined in the model of interests developed in Chapter 1 come into play most visibly. Conflicts between car workers, consumers and shareholders over the relative balance between wages, prices and profits, are likely to dissolve in the face of calls for a radical rethink about the use of cars. They are all likely to try to resist change. And yet, everyone has to breathe, and consumers are beginning to ask for improvements, as are governments. Even so, any really radical changes in this field will have to be extensively negotiated, which perhaps is why technical fixes are popular, since they seem to avoid the need to face up to major changes.

Energy supply fixes

The motor car represents a key consumer product and as such, in principle, consumers can influence the nature of the technology by their purchasing decisions. If enough of them want catalytic converters or electric cars, eventually this demand will be met. However, as noted above, this is not the case when it comes to the basic technology used for energy generation.

Power plants are large capital items, designed, funded, built and operated by companies who are relatively remote from direct consumer influence – it is just the electricity that consumers buy. Nevertheless, changes in design do occur, in part due to the imposition by governments of new environmental regulation, and, so far to a lesser extent, the growth of consumer interest in green power retail schemes.

A full-scale switch over to renewable energy is not the only option for energy generation and supply companies. There are a range of technical fixes for conventional power plants. For example, one way to reduce emissions from fossil-fuelled power stations is to switch fuels: as has already been noted, burning natural gas instead of coal or oil produces less carbon dioxide per kWh generated. Gas burning has become an increasingly popular option for electricity generation in recent years, not just because of environmental concerns, but because gas is relatively cheap and combined cycle gas turbines (CCGT) are very efficient. In the UK, following the privatisation of the electricity industry in 1990, around 20 GW of CCGT capacity has been installed or planned – representing around a third of the UK's electricity generation capacity. Similar levels of expansion are occurring elsewhere.

Even better, there is the option of making use of the waste heat, by designing plants to operate in the so-called combined heat and power (CHP) mode. As noted in Chapter 2, this can more than double the overall efficiency of energy conversion – and so halve the amount of carbon dioxide produced per kWh of total energy generated. Indeed efficiencies of up to 80 per cent or more have been claimed – far higher than that for CCGTs. The economics can however be less straightforward than for CCGTs, since optimal operation of CHP plants requires access to a suitable heat load, and varying demand for heat is usually less easy to deal with than variable demand for electricity. Nevertheless, gas-fired CHP, or 'co-generation' as it is sometimes called in the USA, is becoming increasingly popular around the world. It can be used at various scales, right down to micro CHP plants in individual buildings. The UK has a target of installing 10 GW of CHP capacity by 2010.

However, these power plant fixes, whether CCGT or CHP, still rely on using natural gas as the fuel, and switching to natural gas is only a partial solution to the environmental and resource problem. As we have seen, the world's gas reserves will not last forever, and burning it still generates some carbon dioxide. The gas resource issue seems likely to be addressed in the interim in the UK by importing gas from Russia and elsewhere – although reliance on a gas pipeline crossing some of the more politically unstable areas of Eastern and Central Europe raises serious issues in terms of maintaining security of supply. Similar issues will have to be faced elsewhere in the world if, as seems likely, gas is to be a major interim energy source for electricity generation.

Given that around the world coal reserves are more extensive that those for gas, there has been some interest in trying to clean up emissions from coal combustion. There are two main technologies. Fluidized bed coal combustion plants are more efficient in burning coal than conventional coal plants and can reduce acid emissions. With Integrated Gasification Combined Cycle technology (IGCC), coal is converted into gas which is then burnt in a CCGT. However, these as yet only partially developed technologies are likely to be expensive, with capital costs per kW for fluidized bed plants perhaps twice that for CCGT, and even more in the case of IGCC. Moreover, the overall energy conversion efficiencies of these plants seem likely to be lower than for gas-fired CCGT plants, possibly only around 40 per cent compared with 55 per cent and higher for CCGTs. So IGCCs will produce more carbon dioxide per kWh generated than CCGTs, and of course even more than CHP.

There is one final technical fix on the supply side that deserves mention – carbon *sequestration*, or storage of carbon dioxide. Whether it is coal, oil or gas that is burnt, the problems of dealing with the resultant carbon dioxide remains. So why not capture it and then store it? We saw earlier that trees can absorb carbon dioxide from the atmosphere and store it at least for as long as they survive, and other forms of biomass can perform the same function. Indeed it has been claimed that changes in agricultural practices, including a 'no-till' policy, can absorb large amounts of carbon dioxide. Between them sequestration in soil and trees might absorb perhaps 15 per cent of the global emissions (Royal Society 2001).

However, there is another more direct, but more expensive, option, based on the idea of collecting the gas directly as it is emitted and then storing it as a gas or as a liquid. Clearly collection is not realistic for most vehicle exhausts, but it might be possible for power stations. But, although the technology may well improve, it is far from cheap to capture carbon dioxide that would otherwise vent up the power station chimney. Estimates suggested that it might add 50–80 per cent to the cost of electricity, and could reduce the efficiency of energy conversion by around 10 per cent. Moreover, finding somewhere to store the large volumes of gases is far from easy. The most promising option is to store it as compressed gas in depleted oil and gas wells.

There is a symmetry in this 'carbon sequestration' concept, since that is where at least some of the carbon originally came from. There are also situations where carbon dioxide gas injection can be used to improve the efficiency of gas or oil extraction before the wells are empty. However, overall there may only be space in empty wells for a few decades of emissions. Estimates of global storage capacity range from 40–100 giga tonnes for oil wells and 90–400 giga tonnes for gas wells. Given that in 1999 some 6 giga tonnes of carbon dioxide were emitted by power stations world-wide, and emission levels are increasing, this storage capacity could only be sufficient for at most 80 years' worth of emissions.

There could however be much more space in some saline aquifers, although there are relatively few 'closed' or sealed aquifers suitable for secure storage, possibly sufficient for around 50 giga tonnes. By contrast there is much more room with less secure 'open' aquifers, possibly enough for up to 13,000 giga tonnes, or around 2,000 years' worth of current emissions. However, there is the problem of ensuring that the gas does not escape at some point due to geological shifts and there is a range of other environmental unknowns. For example, sudden carbon dioxide emissions on a large scale could have massive environmental impacts. A sudden release of carbon dioxide from naturally formed gas in Lake Nyos in Cameroon in 1986 killed over 1,500 people by asphyxiation. That could pale into insignificance compared with what would happen, assuming we went ahead with large-scale sequestration, if a major aquifer store were to be breached. Huge amounts of gas could be released, potentially several years' output from several power plants, which could lead to a runaway greenhouse effect.

There is also the newly emerging, but also very speculative, option of ocean sequestration. In addition to the idea of pumping carbon dioxide into the depths, there is some interest in the idea of enhancing sequestration in the oceans by seeding areas with iron oxide. However, modifying the ecology of large areas of the sea could create a range of environmental problems and this approach has been seen as likely to be expensive and inefficient.

In the end, despite its attractions, carbon sequestration, by whichever means, might be seen as a rather inelegant approach to dealing with the problem of emissions – essentially trying to deal with the problem after the event. Moreover, it may not even be a permanent solution, since the gas may be released at some point. Surely it would be better not to produce so much carbon dioxide in the first place. One way to achieve that is by trying to reduce demand for energy, by adopting more efficient energy technologies at the point of use. We have already looked briefly at some options open to domestic consumers for using energy more efficiently, but we will now look at energy conservation in more detail.

Energy conservation

Energy conservation at the point of use is arguably a much more effective and longer term technical fix for environmental problems than trying to collect the emissions: cutting demand, and therefore reducing pollution at source. It can also be more effective than switching over to cleaner types of energy generation. In general, investing in end-use energy efficiency is usually much more cost effective than investing in new supply technology, and the potential for energy (and therefore carbon dioxide) savings in most domestic, industrial and commercial end-use sectors is considerable (PIU 2002).

This is not surprising given that the issue of energy waste has only been considered a significant problem in recent years, and most energy-using appliances have until recently been designed as if energy was almost free. Now however, technologies for cutting energy waste are emerging across the board. Industry has already made great strides at developing more energy-efficient production systems – not least as a way of cutting costs. But returning to the more familiar consumer product sector, there are low-energy refrigerators, cookers, TVs, computers and so on, which typically use a fraction of the power of traditional devices, often at only slightly more purchase cost – and in use they save consumers money.

Energy-efficient lighting is currently a major area of interest. Low-energy fluorescent light can use 70 per cent less power than conventional incandescent bulbs, and although they cost more, they last longer and can soon pay back the extra cost in the energy savings they make.

Table 4.1 gives an idea of the relative cost effectiveness of some key domestic energy-efficiency options, in terms of the time it takes to pay back the original cash outlay from the savings in fuel costs. As can be seen, most of the energy-saving techniques are much more cost effective than domestic-level energy generation technologies like solar heat collectors. Photovoltaics are even worse: at present they are very expensive with, consequently, long payback times.

As Table 4.1 shows, many of the best energy conservation measures in cost terms are related to the basic fabric of houses, and energy-efficient house design is a key area of development. This is not surprising since residential buildings currently consume around a third of the total energy used in many developed countries.

Clearly, not all the options are equally attractive – so the right mix needs to be chosen. That is also true in a more general sense. There is a risk of just seeking technical fixes that initially appear attractive, but are based on complex technologies that may not be viable in the longer term. For example, a study of the three major Low Energy Demonstration housing projects in the new city of Milton Keynes in the UK found that many of the more complex energy-saving devices (including heat pumps and heat exchanger systems) that had been added on had failed, often since they were not fully understood by the householders or were poorly maintained

Table 4.1 *Simple payback periods (UK) for retrofitting domestic measures (ranked in order of payback times)*

Renewable or efficiency measure	Typical domestic savings (£)	Simple payback (years)
Hot water tank jacket	10 to 15	0.5 to 3
Central heating programme/thermostat	15 to 40	2 to 6
Low-energy lighting (per compact fluorescent light)	2 to 5	1 to 3
Loft insulation to 200 mm	60 to 100	3 to 5
Cavity wall insulation	65 to 140	4 to 8
Draught-proofing	0 to 20	6 to 20
Condensing boiler (full cost gas CH)	20 to 200	6 to 12
Marginal cost when replacing boiler	50 to 80	3 to 5
Solar hot water heating system	20 to 100	12 to 40
External wall insulation	60 to 140	20 to 35
Low emissivity double glazing (full cost)	40 to 100	60 to 110
Photovoltaic roof tiles	90 to 180	70 to 135*

*Before government grants (which cover 50% of installation costs)
Source: Ian Bryne, National Energy Foundation, paper to the UK Solar Energy Society Conference 'Renewable energy: what can it do for the environment?', 23 February 1995. Data from National Home Energy Rating Scheme/NEF Data; Energy Saving Trust; IEA CADDET case studies. Updated by Ian Byrne for this book, October 2002

by the suppliers. The conclusion was that, in general, rather than adding complex gadgets, which could go wrong and needed careful optimisation to operate effectively, it was better to design in passive systems into the building framework. Integral passive solar systems and attention to the basic energy-efficient design of the buildings were thus seen as far more robust options (Dougan 1998).

There are many examples of well-designed, low-energy houses around the world. For example, there are now some dwellings in North America, Scandinavia, Germany and Switzerland that use about 80 per cent less energy than conventional new homes and about 90 per cent less energy than existing homes. In northern climates, the energy consumption for space heating can be reduced by over 95 per cent. This has been achieved by technology that was first demonstrated in Sweden and Canada fifteen years ago, and which came to be called super insulation. Super-insulated buildings have very high insulation levels, energy-efficient windows, draught-proof construction, and a mechanical ventilation system that recovers heat from the exhaust air to pre-heat the fresh air in winter. There are now several hundred thousand such buildings in Scandinavia and North America, mainly with timber-frame construction. They are also increasingly common on mainland Europe (Olivier 1996).

So it seems that given proper attention to design and insulation, houses can run with almost no need for external power supplies other than for cooking and lighting. There are many exotic-looking solar houses, but Figure 4.1 shows a plan for a conventional-looking house, which its designers say would nevertheless be a 'zero-fossil fuel, zero CO_2 house'. It is very heavily insulated, and makes use of solar heat via a passive solar conservatory for space heating, and via roof-mounted active solar collectors for hot water. The small requirement for electricity for its low-energy lighting and electrical appliances is met, when sunlight is available, by integral solar photovoltaic cells on the roof: at other times power is taken in from the national grid, but this would be balanced by exporting excess electricity during sunny periods.

Similar developments are occurring in the non-residential buildings sector – in offices and factories, schools and other public buildings – with heat recovery systems, low-energy lighting and improved insulation all playing a part.

It is not just the houses, offices and their energy-related equipment that can be improved in terms of energy efficiency. So can the various energy-using devices they contain. In addition to low-energy lights, a

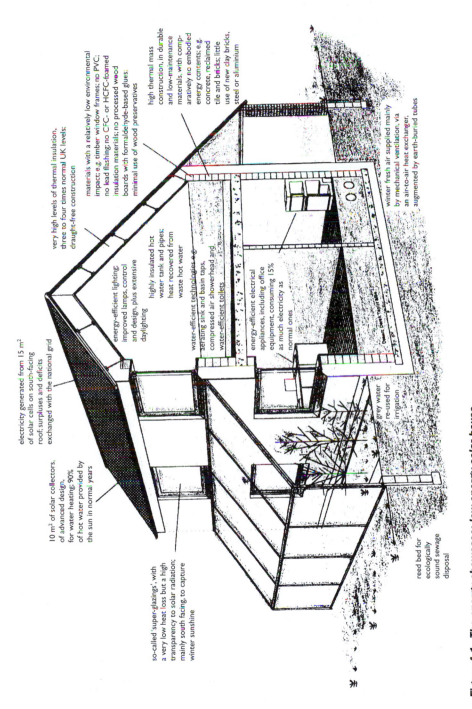

very high levels of thermal insulation, three to four times normal UK levels; draught-free construction

materials with a relatively low environmental impact; e.g. timber window frames; no PVC; no lead flashing; no CFC- or HCFC-foamed insulation materials; no processed wood boards with formaldehyde-based glues; minimal use of wood preservatives

high thermal mass construction, in durable and low-maintenance materials, with comparatively no embodied energy contents; e.g. concrete, reclaimed tile and bricks; little use of new clay bricks, steel or aluminium

electricity generated from 15 m² of solar cells on south-facing roof; surpluses and deficits exchanged with the national grid

energy-efficient lighting; improved lamps, control and design, plus extensive daylighting

highly insulated hot water tank and pipes; heat recovered from waste hot water

winter fresh air supplied mainly by mechanical ventilation, via an air-to-air heat exchanger, augmented by earth-buried tubes

water-efficient technologies e.g. aerating sink and basin taps, compressed air showerhead and water-efficient toilets

energy-efficient electrical appliances, including office equipment, consuming 15% as much electricity as normal ones

10 m² of solar collectors, of advanced design, for water heating; 90% of hot water provided by the sun in normal years

grey water re-used for irrigation

so-called 'super-glazings', with a very low heat loss but a high transparency to solar radiation; mainly south facing to capture winter sunshine

reed bed for ecologically sound sewage disposal

Figure 4.1 The energy showcase: low-energy solar house

(*Source:* D. Olivier, Energy Advisory Associates, UK, http://www.EnergyAdvisoryAssociates.co.uk)

wide range of cookers, freezers, washing machines and so on are emerging which use less energy than earlier models. Similarly for the various electronic devices that now fill our homes – computers, videos, DVD players and the like.

However, there is one area where substantial gains have still to be made. Most domestic and office equipment still consumes some power when it is ostensibly switched off, or in 'stand-by' mode. Some of this is because they have LED lights or timers that remain on, but most is due to the fact that many electronic devices actually run at low voltages and have a transformer to convert mains power (at 220–40 volts in Europe, 110 volts in the USA) to the necessary operating voltage (often 12 volts). The transformers consume a significant amount of power and, in some cases, remain in operation even when the device is not being used, as you can tell, since it gets warm. This is a very poor way to heat your home or office. Nationally, this hidden power drain can be the equivalent to the output of several large power stations.

In the USA these losses are known as 'energy vampires', and 'vampire slaying' has become a priority. This is not surprising given that it has been estimated (by Cornell University) that vampire losses just from timers, clocks, memory and remote on and off switches, costs the USA $3 billion p.a. and used the output of around seven power plants. When all the other losses are added in, the scale of the energy waste is even larger. For example, in a speech to the Department of Energy in 2001, in which he evidently was the first to use the term, President Bush suggested that the overall vampire losses associated with computers, televisions and other common household and business appliances cost the country 52 billion kilowatt hours of power annually, the equivalent output of twenty-six power plants.

The limits of conservation

Energy conservation at the point of use clearly has significant implications. However, as with all technical fixes, there are limits and there can be some implementation and operational problems with energy conservation systems.

1 In relation to energy-efficient appliances, there is likely to be a delay in deployment, particularly in the domestic sector. Domestic consumers only replace appliances occasionally, when they are old or broken, so it

would take time to replace the existing range of equipment with more efficient systems.

2 While over their lifetime the more efficient appliances will cost consumers less to run, they may cost more to buy, and this can be a disincentive for those with limited budgets. The old age pensioner in a damp, draughty flat may be able afford a cheap one-bar electric fire but is unlikely to be able to afford to invest in a more efficient central-heating system.

3 House owners who, due to pressure on jobs and careers, tend these days to move regularly, may not think it worthwhile to retrofit energy-saving measures like insulation in their home. The payback time is too long.

4 In addition to these 'social' problems, there can be technical problems. For example, houses designed with very good insulation and air-tight double/secondary-glazed widows can be stuffy and there can be in some areas a build up of mildly radioactive radon gas inside, emitted from rocks under the building. In which case some form of powered ventilation may have to be provided – using energy. Ventilation may also have to be provided to avoid condensation.

Designers and technologists can try to minimise and deal with problems like this and grant aid schemes can be instituted to enable disadvantaged groups to install energy-efficient appliances, but it seems clear that, while technology can resolve some problems, it also sometimes creates further technical, and in some cases social, problems.

Reducing energy demand

Despite the problems discussed above, the potential for technical fixes in the conservation field seem very large, especially if the necessary institutional support structures exist. There is a need for advice agencies to provide information to consumers and, also, possibly, in some cases, for grant schemes, to stimulate uptake. There is also a need for economic assessment frameworks to allow more realistic comparisons to be made between the benefits of investing in energy-saving technologies as opposed to investing in new supply technologies.

Increasingly it is being realised that managing energy demand is just as important as managing energy supply, and some new 'demand management' technologies have emerged that may help. For example, some consumer devices, such as freezers, do not need power continuously. They can be disconnected for a few hours during peak demand energy times, thus saving consumers' money. Electronic 'smart plug' devices have been developed which cut off power from selected devices during peak demand periods when the price of electricity is high. There are countless technical fixes of this sort which could help reduce demand without reducing utility.

As has been indicated earlier, historically the emphasis has been on developing new generation systems: by contrast much less attention has been paid to energy saving and reduction in waste. These priorities clearly must be reversed. It is conceivable that overall energy demand might be reduced by up to 70 per cent or more in many sectors, given time. That would make the problem of finding sustainable supplies of energy much easier and, in terms of our criteria, investing in energy efficiency obviously represents a key element in any realistic sustainable energy strategy (Olivier 1996).

That said, as we have seen, there can be problems with the implementation of energy conservation measures. Some are technical problems but some are more fundamental, raising questions about our whole approach to energy use.

Perhaps the most serious problem is what is known as the 'rebound effect', which, as we will see, could undermine the effectiveness of energy-saving measures. To put it simply, the money that domestic consumers save by adopting energy conservation measures may be spent firstly on maintaining higher temperature levels for comfort, and then on other energy-intensive goods and services, like foreign holidays by plane. The result can be that at least some of the initial energy and emissions savings may be cancelled out. The exact scale of this rebound effect is debated, with optimists suggesting that, since most products and services are less energy intensive than direct energy use, the rebound effect associated with extra spending on goods and services might only account for a 10 per cent reduction in emission savings that would otherwise have been achieved (Herring 1999).

Nevertheless, overall, it does seem likely that, all other things being equal, if any commodity becomes cheaper, then more of it is used. The optimists sometimes argue that there may be saturation levels for consumer demand in the affluent industrial countries, so that energy use may not continue to grow. Any new wealth will be spent on non- or low-energy consuming services, or on more energy-efficient consumer technology. However, so far, at the macro-economic level, energy use by consumers in affluent countries shows no sign of decreasing, despite dramatic increases in energy efficiency. Moreover, as more people around the world join in the race to material affluence, demand seems certain to continue to increase.

This is not to suggest that energy conservation is not vital, since it is foolish to waste energy that has been produced at such a large potential environmental cost, and, in general, increased efficiency should make for economic competitiveness. However, as pressure for an improved quality of life from an expanding world population grows, the increased energy demand could outpace the increase in savings that can be made, even given major commitments to improved energy efficiency. It may be that, at best, energy conservation via the adoption of energy-efficiency measures and demand management will only be able to slow down the rate of increase in global carbon emissions. To actually reduce it requires a switch to non-fossil fuels.

Conclusion: sustainable energy fixes?

Some technical fixes can be useful and important, but others often seem to offer only partial and short-term solutions. For example, switching to cleaner fossil fuels can help reduce emissions, so may carbon sequestration, but the only long-term solution to global warming is to stop burning fossil fuels and find other more sustainable ways of meeting energy needs. It is the latter options we now need to explore in more detail, that is, the more radical, longer term options that promise more sustainable solutions. You might call them **sustainable energy fixes**.

Energy conservation is an obvious priority. We have seen that there is a range of technical fixes in the domestic appliance and housing fields and in most other sectors that can help reduce energy demand, but there would still be a need to generate power somehow. Conservation, while vital, does not eliminate the need to find new clean and sustainable sources of energy.

Therefore the basic issue we must now explore is as follows: what are the longer term non-fossil energy supply options open to us? The options in terms of energy resources are relatively clear. Figure 4.2 illustrates the basic energy incomes and sources this planet enjoys. Some of them hold out the prospect of supplying power for the long term – the key criterion for sustainability developed in Chapter 2.

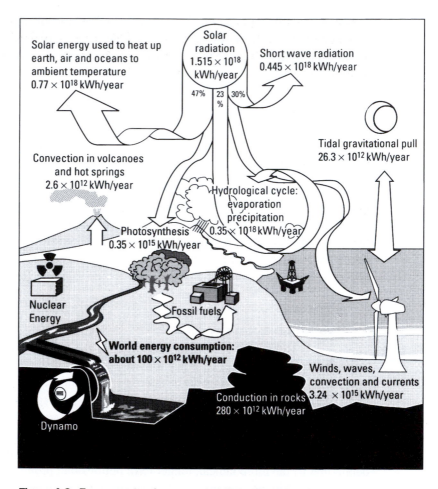

Figure 4.2 *Energy: natural sources and flows (in kWh/yr)*
(*Source: Physics Review* 2(5), May 1993: 8. Reproduced by permission of Philip Alan Updates. Data from G. Boyle and P. Harper, *Radical Technology*, Wildwood House, London, 1977)

As can be seen from Figure 4.2, most of the energy sources derive directly or indirectly from the sun. Fossil fuels are really just stored solar energy, trapped over the millennia underground as coal, oil and gas. However, once these reserves have been used up, they are gone forever. By contrast most of the so-called renewable sources are based on contemporary solar inputs, providing direct heating or producing winds, waves, water flows in rivers or heads of water in lakes, or short-term stored solar energy in the form of **biomass** – plant life that can be used as a fuel.

The only non-solar options are tidal power, which is the result primarily of the gravitational force of the moon acting on the seas, and the geothermal energy deep in the earth, and nuclear energy, from the break-up of the nucleus of atoms. Radioactive decay from natural nuclear processes within the earth is what supplies some of the geothermal heat, but nuclear heat has also been released artificially, by nuclear fission and nuclear fusion. Nuclear sources are not indefinitely sustainable, but their proponents argue that some of them they can nevertheless supply power long into the future.

To summarise, if more radical and sustainable technical fixes in relation to new energy supplies are required, the only non-fossil options known at present are nuclear fission, nuclear fusion and the various forms of renewable energy. The next chapters look at these options in detail, starting first with nuclear power.

Summary points

- Technical fixes and product redesign can help reduce environmental impacts, for example by improving energy efficiency.

- In the energy sector, energy conservation and the more efficient use of fossil fuels are obvious priorities, although there can be problems, including the rebound effect.

- However, if full sustainability is to be achieved then there may be a need for more radical technical fixes.

- There are a range of potentially sustainable energy fixes including renewable energy technologies and nuclear power.

Further reading

A good starting point to explore the idea of technical fixes is *Factor Four* (Earthscan, London, 1997) in which Amory Lovins and his co-authors explore options for 'doubling wealth and halving resource use'. Lovins *et al.* take their argument further in the subsequent book *Natural Capital* (Earthscan, London, 1999).

Amory Lovins, from the Rocky Mountain Institute in Colorado (see the list of contacts in Appendix II), is an energetic exponent of improved energy efficiency; see his seminal *Soft Energy Paths* (Penguin, London, 1977).

For an introduction to technical fixes in terms of pollution control and emission reduction from power stations, see the 'Remedies' chapter of the OU text *Power for a Sustainable Future: Energy Systems and Sustainability*, G. Boyle *et al.* (eds) (Oxford University Press, Oxford, 2003), and also the Foresight report 'Power without pollution', produced by the Office of Science and Technology in the UK Department of Trade and Industry in 2002.

Despite the problems discussed in this chapter, carbon sequestration is increasingly seen as one way to deal with carbon emissions. For overviews see H. Herzog, B. Eliasson and O. Kaarstad (2000) 'Capturing greenhouse gases', *Scientific American*, 54–61 and E.A. Parson and D.W. Keith (1998) 'Fossil fuels without CO_2 emissions', *Science*, 282: 1053–4.

For an introduction to green product design see J. Birkeland, *Design for Sustainability* (Earthscan, London, 2002) and H. Lewis *et al.*, *Design + Environment: a Global Guide to Designing Greener Goods* (Greenleaf, Sheffield, 2001).

For an interesting guide to developments in clean car technology see Jim Motavalli's *Forward Drive* (Earthscan, London, 2002).

For a good textbook on energy conservation see Clive Beggs' *Energy: Management, Supply and Conservation* (Butterworth, London, 2002).

5 The nuclear alternative

- Nuclear fission
- Nuclear accidents
- A nuclear decline?

Nuclear power provides nearly 6 per cent of the world's primary energy at present. This chapter looks at the basics of nuclear power and at how it developed up to the turn of the century. The message at that point seemed to be that it was likely to decline as a major energy source due to the economic and safety problems that had emerged. We look at them in some detail. The next chapter then asks, given that its use does not generate carbon dioxide, can nuclear power can be relied on to play an increasing role in combating global warming, or whether, as many environmentalists would prefer, it should be phased out?

Nuclear power

The basis of the human use of nuclear energy is the fact that the nuclei of the atoms of some naturally radioactive materials found in the earth, most notably one of the constituents of uranium, can be induced to split up or undergo **nuclear fission**, a process that releases heat energy and radiation, together with some sub-atomic particles.

Most of the uranium found in the ground is uranium-238 – this designation referring to the number of basic sub-atomic particles in the atomic nucleus. But in natural uranium there is also a small component (less than 1 per cent) of a slightly different variant of uranium, with a slightly different nuclear make-up. It is uranium-235, and as the name indicates, it has three fewer sub-nuclear particles. Variants like this are called 'isotopes'.

U-235 is naturally radioactive. If this small U-235 fraction is separated out from the U-238, and sufficient of it is concentrated in a small space in a so-called 'critical mass', then the fission process can spread and multiply. The reason is that each fission process produces nuclear particles – neutrons – which, if retained within the lump of uranium, can cause further fissions, so that a self-sustaining chain reaction can be established. As a result, large amounts of heat can be produced, along with nuclear radiation and a range of radioactive fragments (see Box 5.1 for further discussion of these emissions and their implications).

The fission process can be controlled so as to be stable and the heat generated can be used to boil water to raise steam, as in a conventional fossil fuel-fired power plant. If the 'fission' or splitting-up process is made to happen very rapidly then an explosion occurs – the atomic bomb. In conventional nuclear reactors the fuel type and core design is such that this cannot happen, but it is, in theory, an outside chance with the fast breeder reactor.

There is a range of technologies for turning nuclear heat into electricity. In the USA the emphasis has been on water-cooled reactors – in part because the first power reactors were designed for use in submarines and had to be compact. Gas-cooled reactors, as initially developed in the UK, tend to be physically larger, and, it now seems, more expensive. Most of the world's 430 or so reactors are of the water-cooled type, chiefly based on the pressurised water reactor (PWR) initially developed in the USA. Typical modern nuclear reactors range up to around 1,300 megawatts of generating capacity.

Nuclear technology is complex and what follows must necessarily be a simplified account. However, the basic concepts are fairly easy to describe. Figure 5.1 shows, in simplified form, the internal layout of a typical PWR, illustrating how the heat energy is absorbed from the reactor core by a pressurised water-cooling loop. This in turn transfers heat via a heat exchanger to generate steam for power production, as with conventional fossil-fuelled plants.

However, electricity is not the only output. All nuclear reactors also produce *radioactive materials* as an inevitable result of the basic fission process. The most radioactive products of the fission process tend to 'decay', that is, lose their radioactivity, relatively quickly. But some of the

Box 5.1

Impacts of radiation

Some forms of nuclear radiation occur naturally (from radioactive material in rocks and from cosmic rays coming in from space) and the nuclear industry argues that, on average, its activities add only small amounts to what most people already experience, i.e. the so-called 'background' radiation. However, even though great care is taken to avoid exposing people to them, some of the products of man-made nuclear fission are unique, not least in the nature and intensity of the radiation hazard.

In addition to producing heat, the nuclear fission process splits fissile materials like uranium or plutonium into radioactive fission products (various chemical elements of lower atomic mass) and also leads to the emission of very energetic nuclear particles (e.g. neutrons and protons) and electromagnetic radiation (gamma rays – a more powerful form of X-rays). All these products of the fission process can have serious impacts on living material. High-energy nuclear particles or gamma rays can damage the material in cells, particularly young, rapidly growing cells. This is why radiation treatment at low levels is sometimes used to treat some cancerous growths: it kills more of the fast-growing cancer cells than of the normal cells. However, uncontrolled exposure at higher levels can cause cancer, and at very high levels it can kill outright.

Some of the radioactive materials produced by the fission process are not very radioactive, but they have long half-lives, and this combination can be a problem. For example, the radiation from plutonium is relatively weak and can be absorbed by a few layers of human tissue. That means if a person accidentally ingests or breathes in a small speck of plutonium and retains it in their body, its presence cannot be easily detected from outside the body, since the radiation is shielded by the body material. However, cells near to the site of the plutonium are continually irradiated, and eventually cancers may be produced.

Clearly, exposure to all forms of radiation can be dangerous. The main forms of protection are as follows: physical barriers, to absorb the radiation; physical separation by distance; and limiting exposure time. Thus nuclear reactors are surrounded by biological shields to absorb radiation; the length of time any operating staff are exposed to radiation, and the degree of its intensity, are carefully monitored; and reactors are usually sited in remote sites, so as to minimise risks to local people in the event of an accidental release of radioactive material.

initially less radioactive materials can remain active for very long periods. For example, it takes around 24,000 years for the activity in one of the products of fission, plutonium, to fall by a half, the so-called 'half-life'. It would take even longer for the activity level to fall off to negligible levels – say ten half-lives, or around 240,000 years.

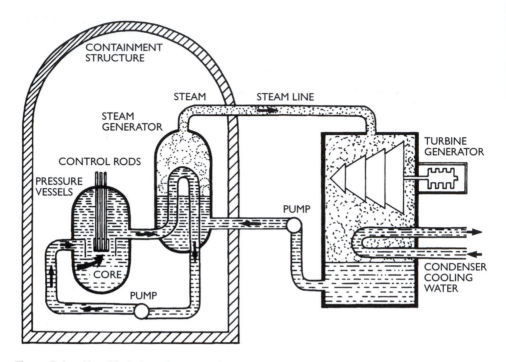

Figure 5.1 *Simplified plan of a pressurised water reactor*
(*Source:* Sally Boyle, Open University, 1996)

Plutonium is created when neutrons emitted from the fission of U-235 atoms are absorbed by some of the left-over U-238 atoms. However much you try to increase the proportion of U-235 in natural uranium (a process know as 'enrichment') there is always some U-238 left – and it is this that gets converted into plutonium, a new, heavier element (its designation is Pu-239).

Plutonium is the key ingredient of nuclear weapons, and, being fissionable, it can also be used to fuel reactors. In particular it can be used in a fast breeder reactor, in which the radiation produced by the fission process converts otherwise 'non-fissile' (i.e. non-fissionable) uranium into plutonium, thus generating or 'breeding' some new fissile fuel. To get access to the plutonium produced by fission processes, the old ('spent') fuel from conventional reactors, or the fuel from breeder reactors, must be chemically treated to separate the plutonium out from the other nuclear wastes. This process is messy and generates secondary wastes. Indeed reprocessing is what generates the bulk of nuclear waste, even if it is at lower levels of radioactivity than the original wastes.

Radioactive materials and the radiation from them can be very dangerous for living things and extreme care must be taken to avoid exposure (see Box 5.1). This poses problems for each stage of the nuclear energy generation process, from the start to the end. Miners involved in extracting uranium from the ground can be exposed to radioactive radon gas, and at the other end of the nuclear fuel cycle, there is the issue of *nuclear waste* produced by reactors. As we shall see in the next chapter, the issue of what to do with it has become increasingly urgent – so far, while there are short-term storage facilities, there are no longer term repositories, although some are planned. The volumes involved may not be very large, compared with those resulting from some other industrial processes, but nuclear waste poses serious containment problems given that, as we have seen, they can remain dangerous for millennia.

In addition to the problem of waste, there is also the risk that radioactive materials, and the radiation from them, can escape during the routine operation of power plants. Extreme care is taken to ensure that emission levels are low. Government controls are very strict and the engineers and technologists involved have tried their best to make the systems foolproof. But no technology can be 100 per cent safe, and accidents can always happen. The chance of major accidents is very low, and can be reduced if sufficient money is spent on safety. But there still remains a chance of accidents, and the impacts can be very severe.

Nuclear accidents

A review of the potential impacts of energy technologies, including nuclear plants, published in the journal *Nuclear Energy* in 1993, noted that if 10 per cent of the radioactive material in the core of a typical PWR was accidently released into the environment, 'total mortality of approximately 10,000 people is expected in a typical Western European location with appropriate counter measures'. It added 'the probability of such accidents is obviously open to speculation. Some industry sources put the probability as low as 1 per 10^8 (100,000,000) GW years for a PWR. Actual world operating experience to date with a wider range of reactor types is closer to 1 in 10^3 (1,000) GW years' (Eyre 1993: 324).

These odds sound reasonably long, even in the worse case quoted above: one chance of a major accident every thousand years for each gigawatt of nuclear plant operating. However, with, say, 100 reactors rated at 1 GW each, the chance of an accident increases significantly – to once in every ten years, on the basis of the 10^3 GW years figure. Statistical chances are unreliable guides to reality, but, so far, there have been three major nuclear accidents – a fire at the Windscale military plutonium production plant in the UK in 1957, the Three Mile Island accident in the USA in 1979, and the Chernobyl accident in 1986. Given that there are now actually around 430 reactors operating around the world with a combined capacity of about 340 GW, we seem so far, to have been statistically fortunate, or perhaps the 1 in 10^3 per GW years figure is a little high.

The fire at the Windscale plutonium production pile in Cumbria (the site is now known as Sellafield) resulted in the release of radioactive material, including around 20,000 curies of radioactive iodine (one curie is defined as the radioactivity, measured in terms of the energy released, of one gramme of radium – the radioactive material first separated out chemically by Madame Curie). Crops over a 300 sq. mile area were destroyed as a precaution and two million litres of milk were poured away (Arnold 1992). The near melt-down accident at Three Mile Island was contained, although there were some releases of radioactive material and at one time mass evacuation of at least 630,000 local people seemed imminent (Garrison 1980). In neither case were there any direct casualties, although there have been suggestions that some early deaths might have occurred subsequently in the population downwind from the Windscale plant (McSorley 1990). The Chernobyl accident, by contrast, was much more severe, and led to concern right across the world as the radioactive plume spread across North-West Europe and even, although much weakened, to North America. The accident occurred during an attempt to check the plant's safety system, when it was running at very low power, and only a proportion of the radioactive material it contained was released into the atmosphere (Read 1993). Even so, quite apart from the thirty-one deaths on site and among firefighters, there have been reports of subsequent deaths and serious illness in the region around the plant. Estimates vary from a few hundred early deaths to several thousand, while some reports suggest that, ultimately, the health of many more people may be adversely affected (McSorley 1990; Medvedev 1990). Box 5.2 attempts to summarise the scale of the health impacts, which were widely seen as very significant.

The nuclear industry clearly faced a major problem. As Harold Bolter, a director of British Nuclear Fuels Ltd (BNFL) until 1994, noted in his very frank book *Inside Sellafield* (Quartet 1996), 'The UK's National Radiological Protection Board has estimated that around 30,000 fatal cancers will occur over the next forty years in the affected parts of Russia and Western Europe as a result of that terrible event.'

However, as is noted in Box 5.2, there have been some more recent reassessments of the impacts, for example by UNSCEAR, the UN Scientific Committee on the Effects of Atomic Radiation. As a result, by December 2000, Hugh Collum, then Chairman of BNFL, speaking at the 5th Nuclear Congress, was able to note that 'The recent UNSCEAR report on the after-effects of Chernobyl, fourteen years on, noted that, apart from a localised and high level – but entirely treatable – incidence of thyroid cancers among children – there has been no evidence of any long term public health impact due to radiation exposure.'

Box 5.2

Chernobyl death estimates

Morbidity and mortality estimates for the Chernobyl accident were produced for an international conference 'One Decade After Chernobyl', organised by the International Atomic Energy Agency, the World Health Organization and the European Commission on the tenth anniversary of the accident (http://www.iaea.org/worldatom/Programmes/Safety/Chernobyl/). In summary, the analysis was as follows.

- During the accident, three workers were killed. Among plant workers and firemen who responded to the event, 237 were suspected of acute radiation sickness (ARS) and admitted to hospitals for observation, where 134 cases of ARS were confirmed. Of those people treated for ARS, 28 have died within 10 years of the accident as a result of their injuries. The report notes that 'the others have survived but are in generally poor health and psychological condition'.
- Subsequent to the initial event, a workforce of approximately 200,000 'liquidators' were hired or ordered to gather and bury radioactive material released by the blast. It was projected that this population of workers will suffer an excess burden of 2,500 cancers as a result of their clean-up work.
- Residents of communities off-site may be expected to suffer an excess burden of 2,500 cancers as a result of exposure to fallout from the accident. Within the first 10 years, approximately 500 cases of thyroid cancer amongst people who were children at the time of the accident have been confirmed, and it was estimated that there could be between 4,000 and 8,000 cases in total in the coming years. However, 90 per cent or more of these cases were thought to be likely to be curable (surgical removal of the thyroid was widely adopted).

Some observers claim that the eventual death toll might be very much higher than these figures suggest. For example, a 1990 study by Medvedev suggested that there could be up to 40,000 early (i.e. premature) deaths (Medvedev 1990). Given that it can take up to 20 years for cancers to emerge, it will still be some while before predictions like this can

continued

be assessed. But certainly some horrific reports have emerged. For example, in a retrospective review of the accident, the *Guardian Weekly* (19 December 2000) noted 'Ukrainian government figures state that more than 4,000 clean-up workers have died and a further 70,000 have been crippled by radiation poisoning. About 3,400,000 people, including 1,260,000 children, are suffering from fallout-related illnesses. Unofficial statistics put the casualty rates much higher.'

However, it has, perhaps rather harshly, been suggested that some of these figures may have been inflated in order to attract aid, and certainly it can be hard to obtain reliable data, given that the break up of the USSR followed soon after the accident at Chernobyl. It also been suggested in a UN report that some of the illnesses subsequently emerged as the result of the stress of over-zealous evacuation and forced relocation, and that some could even be put down to psychosomatic effects (UNDP and UNICEF 2002).

Clearly, something of a revision of 'official' views has been underway. For example, a UN assessment, produced by the UN Scientific Committee on the Effects of Atomic Radiation in 2000, suggested that, apart from the initial deaths and 1,800 cases of (potentially treatable) thyroid cancer in children, there was 'no evidence of a major public health impact' (UNSCEAR 2000).

To follow up this debate see http://www.nea.fr/html/rp/chernobyl/chernobyl-update.pdf and for a summary by the Nuclear Energy Agency of the report by the UN Scientific Committee on the Effects of Atomic Radiation, produced in 2000. For the full UNSCEAR report see http://www.unscear.org/pdffiles/gareport.pdf and in particular, the detailed annex at http://www.unscear.org/pdffiles/annexj.pdf

Subsequently, John Ritch, Director General of the World Nuclear Association, in an interview on BBC World TV said that the UNSCEAR report indicated that 'the total number of people who died at Chernobyl is fewer than 40. This includes the consequences from the accident'. He added that UNSCEAR believed that there were '1800 thyroid cancer cases that resulted from the accident – but those resulted in only two, or three, or four deaths' ('Hardtalk', 9 July 2001).

Views like this have not gone unchallenged. For example, responding to press commentary that had claimed that the longer term health impact of Chernobyl was low, Patrick Gray, one of the UN assessment team involved with the study of Chernobyl impacts (at that point still in draft form) argued that

> the suggestion that the only health consequences are likely to be 41 deaths from radiation sickness, together with cases of childhood thyroid cancer 'totalling 1,800 in all', is indefensible. In fact the draft report says that on conservative estimates, a further 8,000 cases of thyroid cancer can be expected.
>
> (*Observer* 13 January 2001)

Even more alarmingly, a Ukrainian Government commission on radiation security has claimed that, even more than a decade after the accident, 24 per cent of babies born near Chernobyl had birth defects.

It is hard to assess such claims, especially when other reports come to different conclusions, and the academic debate continues (Gottlober *et al.* 2001). It seems that what initially might seem like a simple, if grim, exercise in body counting, turns out to be a far more complex and potentially conflict-laden activity, fraught with statistical and conceptual problems, and disagreements over the interpretation of data will no doubt continue. Obviously, the figures for death and health impacts do have to be put in context. Major dam failures can also kill many people, and supporters of nuclear power often claim that, compared with its alternatives, nuclear power is a relatively safe option. For example the Uranium Information Centre claims that 4,000 people were killed due to dam failures around the world during 1970–92. Moreover, it is not just a matter of major accidents. Significant and catastrophic though major nuclear accidents may be for those involved, statistically the total health impact of occasional large accidents can be less than, for example, the cumulative and continuing impact of gaseous emissions from coal-fired plants.

To be fair, there are also concerns about the risks associated with the routine operation of nuclear plants. These are usually considered to be relatively small, especially for the general public, given the tight regulation of the industry. However, these controls may not be equally effective in all countries, and accidents do happen even in the most controlled environments. Moreover, quite apart from the impacts of accidental leaks of radioactive materials, some critics argue that the scale and the health impact of the low-level routine emissions allowed from nuclear facilities has been underestimated (see Box 5.3).

This is not the place to enter into the complexities of statistical 'risk analysis'. Suffice it to say that fossil fuels, coal in particular, clearly can have very large health impacts, as is suggested by the figures in Table 5.1. The most striking thing in this table is the very large number of off-site public injuries and deaths per GWyr associated with coal and oil compared with gas and nuclear fuel. These figures reflect the significant impacts on public health (for example in terms of respiratory diseases) of the acid and other emissions from coal and oil burning, and the lower impacts of gas and nuclear fuel.

However, not everyone has been happy with the data used in this analysis, which was produced by Professor William Nordhaus, a proponent of

Box 5.3

Low-level radiation

The nuclear industry argues that levels of exposure to man-made radiation sources, averaged out across the community, are very small compared with natural levels of exposure. Certainly, the public's exposure to radiation from the routine discharges from nuclear plants is said to be very low (averaging out across the population at 0.1 per cent of the public's typical total exposure), compared with natural sources (representing around 87 per cent of the total average public exposure), in particular radon gas from the rocks in some parts of the country, and compared to occasional medical X-rays (around 12 per cent). However, averaging out exposures over the whole population is perhaps a little dubious. People living near nuclear facilities may well receive much larger doses from routine discharges.

In addition, some discharges are decidedly not routine. For example, the accidental flushing out of around 4,500 curies of radioactive material from Sellafield in Cumbria in 1983 led to the closure of beach areas for many miles around the plant's outflow for six months, as a precautionary measure. Similar problems have occurred at Dounreay in Scotland, where 'hot spots' of plutonium were discovered on the beach. Events like this have occurred all over the world, including at the Cap de la Hague reprocessing plant in France and the Handford nuclear facility in the USA.

It is to be hoped that the precautions taken once a leak is discovered will minimise any heath risks, and, statistically, these leaks may not significantly affect the average risk across the entire population. Even so, that way of presenting the data may not build confidence amongst the public, especially those who might be at higher risk by being near a potential source. More generally, the calculated voluntary risks associated with, for example, exposure to X-rays, or to excess cosmic radiation while flying, are seen as somewhat different in kind from the involuntary risk, for example, of ingesting or breathing in particles of plutonium that have been released into the environment. Many people clearly place a high value on being able to choose the risks they face. Of course, there is also the issue of whether people know about either of these types of risk. Access to the necessary information, and the ability to analyse it, are an important prerequisite for constructive debates and informed responses. However, even with reasonable access to information, there can still be basic disagreements based on values. For example, some people believe that any extra radiation exposure ought to be avoided. Whether that is technically or economically realistic is of course another issue, but values like this do shape responses to nuclear power and cannot be wished away.

For further discussion see the arguments presented by the Low Level Radiation Campaign at http://www.llrc.org.

Table 5.1 *Comparison of estimated fatalities associated with energy production*

	Comparative risk by electricity production by fuel cycle*			
	Occupational		Public (off-site)	
	Fatal	Non-fatal	Fatal	Non-fatal
Coal	0.2–4.3	63	2.1–7.0	2,018
Oil	0.2–1.4	30	2.0–6.1	2,000
Gas	0.1–1.0	15	0.2–0.4	15
Nuclear (LWR[†])	0.1–0.9	15	0.006–0.2	16

* Accidents and diseases per GWyr – including entire fuel cycle, excluding severe accidents
[†] LWR = Light water reactor, the generic term for reactors using ordinary water for cooling, like PWRs
Source: From William Nordhaus, *The Swedish Nuclear Dilemma: Energy and the Environment,* Resources for the Future, Washington, DC, 1997

nuclear power, especially since it excludes 'severe accidents', which, given Chernobyl, might be thought to undermine its credibility. Depending on what figure you adopt for Chernobyl impacts, the comparisons could look somewhat different. Nevertheless, it seems clear that coal and oil are going to have large impacts.

What about the renewables? While, as we have seen, large hydro may have its problems, there is as yet less data on the new renewable technologies. However, it is hard to see how, for example, wind turbines can present the same risks as nuclear plants. By 2002, with around 31,000 MW of wind capacity installed around the world, there had been twenty-one operator deaths, mostly due to falls and injury by blades, and no injuries to the public. By comparison, depending on one's conclusions about the debate discussed in Box 5.2, the figures for deaths per kWh produced are much higher for nuclear plants (with currently some 350 GW installed worldwide) and their associated systems, including uranium mining and waste processing and disposal, which are problems not shared by the renewable energy sources.

It also seems that, quite apart from the quantitative data on health impacts, nuclear risks, from accidents and leaks, are widely perceived as different in kind from other types of risks, not least because they are perceived as involuntary and can lead to illness many years later. Certainly, concern over the potential long-term consequences of major accidents has been one reason why nuclear power has been opposed by most environmentalists. Fears about radiation hazards generated in relation to nuclear weapons have always meant that the general public has been nervous about the risks of nuclear technology, and the accidents and leaks have seemed to confirm that there could also be major problems with civil power plants.

A failed option?

Despite the risks, nuclear power might be seen as the ultimate technical fix. Large amounts of energy can be released from small amounts of relatively abundant material. Certainly, in the period after the Second World War it looked very promising: there were even suggestions that nuclear electricity would be 'too cheap to meter'. The reality has proved somewhat different. Fifty years on after very large investment around the world, it still only provides under 6 per cent of the world's primary energy and the cost of the electricity produced remains high. For example, in the UK, following the privatisation of the electricity industry in 1990, electricity consumers had, in effect, to pay a surcharge of around 10 per cent to meet the extra cost of the 20 per cent or so of the UK's electricity that nuclear plants were supplying. Nuclear power was increasingly seen as expensive, and as a failed option.

Certainly, the later part of the twentieth century was a gloomy one for nuclear advocates. Chernobyl had been a major blow to investor confidence. In addition, the relative cost of electricity from the rival energy technologies had fallen. For example, since the 1980s, the average price of electricity produced by coal plants fell by a factor of two in the USA in real (inflation-adjusted) terms. More recently, nuclear power has been seriously challenged by the so-called 'dash for gas' – the rapid adoption of low-cost electricity generation using small, highly efficient gas-fired turbines. They were cheap and quick to build, with capital costs per kW installed of around a third to a quarter of that for a nuclear plant, and gas was also increasingly cheap. So while some existing nuclear plants had, in effect, paid off their construction costs, and, depending on the market context, should be able to generate power reasonably economically, it is generally accepted that using current technology it would not be economically viable in most countries to build new nuclear plants, since they could not compete with gas plants.

Economic problems coupled with continued public opposition and the need for ever more expensive safety controls meant that, in many countries around the world nuclear power programmes, private or state run, were winding down or halted. Nuclear plants provided around 20 per cent of the USA's electricity but the accident at Three Mile Island effectively halted expansion. Denmark decided not to go nuclear, after a national debate during the late 1970s, followed by a referendum, and the Chernobyl disaster led several European countries with existing nuclear programmes to halt further expansion or to develop phase-out plans. For example, Italy decided to abandon nuclear power in 1987, and Sweden decided to phase out its nuclear plants.

Although France still had a major nuclear programme, supplying around 75 per cent of the country's electricity, in 1997, the Socialist administration imposed a moratorium on new nuclear developments. The SPD government elected in 1998 in Germany decided to phase out Germany's nineteen nuclear plants, with the current plan being for a complete phase-out by 2032. In 1999 Belgium decided to phase out its nuclear programme by 2025 and in 2000 Turkey decided not to invest in nuclear power.

While the nuclear programmes kept going in Eastern and Central Europe, and there were plans for expansion in some countries in the Far East, within Western Europe, at the close of the century, the only sign of a new project was a proposal (later confirmed) for a single new plant in Finland.

As a consequence the nuclear industry seemed to be faced with slow decline. In the short term, some new plants were expected to be started in Asia, so that a moderate increase in capacity might follow. But, as old plants were retired around the world and not replaced, and as phase-out programmes began to bite, unless there were changes in policy, the nuclear contribution would begin to fall off after about 2010 (see Figure 5.2). The cut-off point could, on current plans, be even earlier in the UK (see Figure 5.3).

Certainly, the view from the nuclear industry became somewhat less assertive. For example, in a paper to an international conference on energy organized by the World Energy Council (WEC) in May 2000, Steven Kidd from the Uranium Institute (now known as the World Nuclear Association) commented that, although nuclear plants produced around 17 per cent of the world's electricity at present 'given that there are relatively few reactors currently under construction and that some of the early plants are scheduled for closure, further increases in world electricity output will mean that this share is set to fall'. He was not sanguine about opportunities for renewal of nuclear growth, except possibly in China: 'Although the economic turbulence in this

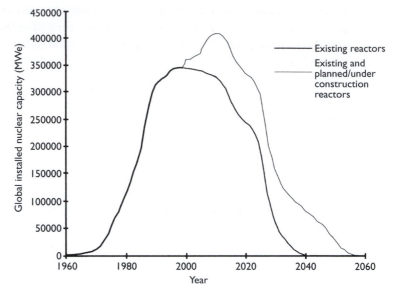

Figure 5.2 *Rise and fall of nuclear capacity worldwide*
(*Source:* Data from Royal Society/Royal Academy of Engineering, 'Nuclear energy – the future climate', 1999)

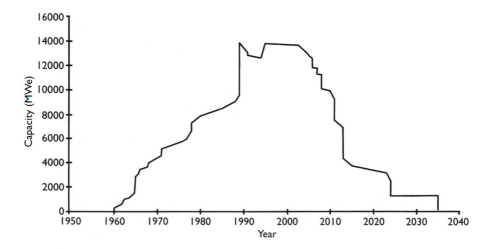

Figure 5.3 *Rise and fall of nuclear capacity in the UK*
(*Source:* Data from Royal Society/Royal Academy of Engineering, 'Nuclear energy – the future climate', 1999)

region has not been helpful, this region appears likely to continue to be the main growth area. In particular, China could feasibly embark on a construction programme similar to that of France in the 1970s and 1980s' (Kidd 2000).

Kidd continued 'Elsewhere the economics of new nuclear plants do not look good, unless one takes a pessimistic view of fossil fuel prices and utilizes a low discount rate. Combined cycle gas turbine plants of similar generating capacity can be built for one third of the capital cost of a nuclear plant, and will also be open several years earlier. This is a powerful set of arguments for private financiers.' He concluded 'hopes for renewed nuclear growth are now largely based upon the connection with the global warming debate on emissions targets'. But he felt this was not a reliable route since it is 'unclear how far various countries will go down the road of carbon taxes, emissions permit trading and the like'.

In another paper to the WEC conference, Hans Holgar Rogner and Lucille Langlois from the International Atomic Energy Agency were also somewhat muted on the prospects of a nuclear renewal based on concerns about climate change.

> It is true that nuclear power offers governments the opportunity to achieve a number of national policy goals, including energy supply security and environmental protection, particularly by reducing air pollution and greenhouse gas emissions. But these policy-related 'benefits' are vulnerable to policy change, and are insufficient by themselves to assure a nuclear future. Similarly, the further internalization of externalities [i.e. the inclusion of environmental costs in the generation costs] – to a large extent already imposed on nuclear power – is a policy decision and it is unclear when and to what extent such policies will be implemented. Those who pin their hopes for nuclear growth on externalities or on the Kyoto Protocol – and ignore reform and the need to innovate – will be doomed to disappointment. The nuclear industry has to bootstrap itself to economic competitiveness by way of accelerated technological development and innovation.
>
> (Rogner and Langlois 2000)

Clearly, these representatives from the nuclear industry were not too confident about the immediate future of nuclear power, even given its potential role, as a non-fossil energy source, in responding to climate change. In the next chapter, we ask whether this pessimism is warranted.

Summary points

- Nuclear power was once seen as a solution to some of the world energy problems – it would be cheaper and cleaner than fossil fuels.

- However, nuclear power is currently not economically competitive with gas-fired generation, and few governments are willing to subsidise it.

- Accidents can happen even with the best designed technologies: although the chances of major nuclear accidents are low, the impacts can be very serious and those that have happened at nuclear plants have led to major public concerns.

- Nuclear operations, particularly those associated with spent fuel reprocessing, involve the permitted routine release of low-level radioactive materials, and many people remain concerned about the impacts, and also about the problem of nuclear waste disposal.

- Concerns about safety, security and economics have led to a decline in enthusiasm for the nuclear option in some, but not all, countries.

Further reading

Walter Patterson's much reprinted *Nuclear Power* (Penguin, Harmondsworth, 1976) remains one of the best general introductions. There are many books and journals on specific aspects of nuclear power, ranging from the technical to the political. For a useful analysis produced by the Energy and Environment programme team at the Royal Institute of International Affairs, which attempts to review the positions taken by each side from a neutral standpoint, see Malcolm Grimston and Peter Beck's 'Civil nuclear energy: fuel for the future or relic of the past' (RIIA, 2000). They produced a follow-up, *Double or Quits*, co-published with Earthscan, London, in 2002.

Web-sites are becoming increasingly central to the debates over these issues. For the UK, see the web-sites run by British Nuclear Fuels Ltd (http://www.bnfl.com) and Greenpeace (http://www.greenpeace.org).

For an international perspective, the World Nuclear Association (previously called the Uranium Institute) also has a very useful site at http://world-nuclear.org. It includes the text of a book on nuclear power (*Nuclear Electricity*, 6th edn, 2000) produced by the Australian Uranium Information Centre.

For an anti-nuclear perspective on global issues see the joint WISE/Nuclear Information and Resources Service web-site at http://www.nirs.org/.

6 A nuclear renaissance?

- Nuclear power and climate change
- New nuclear technology
- Waste management and reprocessing
- A nuclear revival?

In the previous chapter we looked at some of the problems facing nuclear power and saw that the prospects for nuclear power had begun to look rather poor by the end of the twentieth century. However, for good or ill, nuclear power could make a come-back, on the basis of being a non-fossil energy source, with no direct greenhouse gas emissions. In this chapter we look at the new fission technologies emerging and also take a look at nuclear fusion, which some people see as a possible energy option for the longer term.

Nuclear power and global warming

The prospects for a nuclear revival were significantly improved by the turn of the century by the growing concerns about climate change. As we have seen, the climate change issue had first been raised in the early 1990s, but by 2000 it was being taken very seriously around the world. Nuclear reactors do provide a way of generating electricity without producing greenhouse gases such as carbon dioxide, and so around the world there have been increasing signs of interest in the nuclear option as one possible response to global warming and climate change.

Could a renaissance of nuclear power rescue the situation? Perhaps the first point to make is that it is not strictly true that nuclear power does not generate any carbon dioxide. Unlike the other energy technologies, the fuel for nuclear plants has to be extensively processed (from ore) and this is an energy-intensive activity. Although some of the power can come from nuclear reactors, for the moment most will be from fossil fuel plants, which will generate carbon dioxide.

Even so, the complete nuclear power process does produce much less carbon dioxide than fossil plants: as we noted in Chapter 3, one estimate puts the total fuel cycle emissions for nuclear power plants at 8.6 tons per gigawatt hour, compared with 1,058 tons for coal plants.

On this basis, it has been suggested by the UK Atomic Energy Authority that a programme aiming to increase the global nuclear contribution by nearly threefold, up to around 50 per cent of world electricity requirements by 2020, would result in a 30 per cent reduction in global carbon dioxide emissions from what they would otherwise have been (Donaldson *et al.* 1990: 26).

However, there are also some major drawbacks. The most obvious is that all nuclear plants generate dangerous nuclear waste. As has already been indicated, some parts of the nuclear waste remain dangerous for thousands of years. Some countries are trying to develop long-term repositories for high-level waste in remote areas, and there are 'vitrification' techniques for converting some waste into a glassified form, but no one can be 100 per cent sure that the waste can be successfully contained over such long lengths of time. Few communities are willing to accept waste repositories near them, and yet more and more waste is being produced. It is a real problem which many environmentalists feel can only really be tackled if no more waste is created.

Uranium reserves

The next point is that, while some of the nuclear waste will be with us for many thousands of years, nuclear energy generation may only be a relatively short-term option – uranium reserves are not infinite. This is important since there is no point in pushing ahead with new nuclear plants to try to avoid greenhouse gas emissions, if there will not be fuel for them in the longer term.

It is interesting in this context to look back at the early 1990s, when the nuclear industry was trying to promote fast breeder reactors, which could breed plutonium from the otherwise wasted uranium-238 (see Box 6.1). Part of the basis of their campaign was the claim that this would be

Box 6.1

Fast breeder reactors

Fast breeder reactors (FBRs) can in principle extract 50–60 times more energy from uranium fuel, so that the availability of fissile material could, in theory, be extended by up to 50–60 times. Using figures like this, and assuming current uranium reserves are put at 100 years, nuclear enthusiasts sometimes talk in terms of uranium reserves being stretched thereby to 'thousands' of years, although more guarded commentators limit it to around 1,000 years (Eyre 1991).

Certainly, claims that uranium reserves could be extended in practice by a factor of 50 or 60 could be seen to be somewhat optimistic, especially since we are talking about as yet hypothetical systems involving networks of fast breeder reactors and reprocessing plants.

In this context, it is important to realise that it would also take time for a breeder programme to make a significant contribution in energy terms. Despite the name, the breeding process is not 'fast'. In fact it can take years to breed the same amount of plutonium that you started with. The so-called 'doubling time' can be in excess of twenty years, perhaps up to thirty years, especially when the fuel cooling, reprocessing and refabrication processes are taken into account (Mortimer 1990). The word 'fast' simply refers to the type of nuclear interaction that occurs in breeder reactors: it involves fast, high-energy neutrons, rather than the slow, low-energy neutron interaction in conventional reactors.

In a breeder, the fast neutrons convert the non-radioactive part of uranium (U-238) into plutonium, a process that also occurs, but much less efficiently, in conventional reactors. As was noted above, to get access to this plutonium the old 'spent' fuel must be reprocessed, so that a fast breeder-based system would require significantly enlarged reprocessing facilities. Not only would this significantly increase the amount of nuclear waste that was produced, it would also, presumably, increase numbers of shipments of spent fuel and plutonium would have to move between the various reactors and reprocessing plants. That would open up a range of safety and security problems, not least the risk that, despite all the precautions that would no doubt be taken, plutonium could be stolen for bomb-making activities.

So, although it has some attractions, and could extend the lifetime of the uranium resource, there are a range of problems with the fast breeder option. Despite enthusiasm from the nuclear advocates, many of whom at one time claimed that the breeder was the only realistic future for fission, in practice the prospects for breeders are unclear.

Around the world fast breeder projects have been shut down. In the USA, then President Carter was evidently concerned about the potential security problems of plutonium proliferation, and a moratorium was imposed on the fast breeder programme in 1977. In Germany the prototype fast breeder at Kalkar, the scene of major protests by objectors

continued

in the late 1970s, was finally abandoned in 1991. The UK Government was alarmed at
the costs of the FBR project at Dounreay and at the long timescale before a
commercially viable technology might emerge. The FBR programme was halted in
1994. France abandoned its FBR programme in 1997, leaving only Japan with a major
FBR programme, although this has had technical problems, with an accident at Japan's
Monju plant in 1995 involving a sodium fire, which led to a review of Japan's nuclear
programme.

needed if the nuclear industry was to expand significantly since uranium
supplies might be limited. In 1990 the UK Atomic Energy Authority
(UKAEA) suggested that if an attempt was made to expand nuclear power
dramatically on a worldwide basis in response to the threat of global
warming using conventional 'burner' reactors, 'the world's uranium
supplies that are recoverable at a reasonable cost would be unlikely to last
more than about fifty years' (Donaldson et al., 1990: 29).

The UKAEA's journal ATOM was even more specific, carrying an article
in 1990 which suggested that 'for a nuclear contribution that expands
continuously to about 50 per cent of demand, uranium resources are only
adequate for about 45 years' (Donaldson and Betteridge 1990: 19).

Since then, the situation has changed. There have been new finds of
uranium and, given the slowdown of the nuclear programme worldwide,
there is no immediate shortage of uranium; indeed there is something of a
glut, with new finds pushing the resource limit up to at least 100 years
and maybe more, at current use rates. The precise figure will depend not
only on the actual rate of use, but also on assumptions about whether
speculative resources (i.e. anticipated reserves that have yet to be proven)
can be considered as reliable. In practice this comes down to economics.
If prices increase, it becomes worthwhile exploring for and using less
easily available reserves, even if they cost more.

Of course, there must be limits to the economically viable resources.
It might be that, as reserves of easily usable high-quality uranium ore
become scarce, reserves could be stretched by using lower grade uranium
ores. However, the net energy balance might not be favourable: as we
have noted, uranium ore contains only a very small amount of the type of
radioactive uranium needed for fission reactors (the uranium-235 isotope)
and this has to be extracted and enriched to make usable fuel. As just
noted, this is an energy-intensive process. Using lower grade ores would
make it even more energy intensive. At some stage the so-called 'point of
futility' is reached, when, as lower and lower grades of uranium have to

be used, more power is required in order to process the fuel than can be generated from it.

One study suggested that, if a major nuclear programme was launched in response to concerns about global warming, then at least initially more carbon dioxide would be produced net than if the electricity from fossil fuel plants was simply used by consumers directly. The reason is that, although nuclear reactors might eventually provide the energy for nuclear fuel processing, in the initial phase the energy for processing the increasingly lower grade uranium ores would have to come from fossil-fuelled power plants (Mortimer 1990).

There are some other options. If uranium became scarce, some of the world's existing plutonium supplies could perhaps be used as a fuel in conventional reactors. The end of the Cold War has meant that there is less need for plutonium for weapons, and the demise of fast breeder programmes around the world (see Box 6.1) means that there is no other use for the plutonium that is inevitably created by nuclear reactors. A new plutonium-based fuel has been developed called MOX – mixed oxide fuel – this being a mixture of plutonium and uranium oxides. Some people have welcomed this as a way of disposing of plutonium and using bombs for better purposes, although in reality there may be technical problems with using old weapons-grade plutonium in this way and in any case the volumes of weapons-grade material available are relatively small – it might provide the equivalent of one year's worth of uranium supply. More importantly, using MOX in reactors would generate further nuclear wastes, including more plutonium, and there are significant security problems involved with transporting this fuel – it can be used for weapons production. Given the current glut of uranium, and the higher cost of MOX, commercial interest in this fuel is limited.

If and when uranium in the ground did become scarce, so that prices rose significantly, it has been suggested that it might be worthwhile to extract it from sea water, despite the very low concentrations (0.003 parts per million compared to 1,000 ppm, even for low-grade uranium ore). But for the foreseeable future that looks like a very long shot.

New nuclear developments

If we tried to expand nuclear power dramatically in response to climate change, then at some point fuel reserves would become an issue.

However, there are many other problems that would have to be faced long before that. The most obvious are safety and economics – the problems that have already brought about nuclear power's decline.

There are attempts underway to develop new reactor technologies that are safer and cheaper. For example, in an attempt to assuage public concerns over safety, the industry has talked of developing so-called 'safe integral' or 'passive' reactor technology, using small modular units that are designed to be fail-safe (e.g. in terms of emergency cooling) under all conditions.

The search for safer types of nuclear reactor has focused mainly on reactors with passive cooling by natural convection – so that if the cooling system fails there is less chance of the reactor core overheating and the fuel melting down. The industry argues that, since about half the extra capital and operational costs of nuclear plants, over and above that of fossil fuel plants, is the result of the complex safety and control systems, if these can be avoided by the use of passive technology, then plants should be cheaper.

Some, like the Westinghouse AP 600 and AP 1000, are basically upgrades of the original PWR design, with the active control features replaced wherever possible by passive safety systems. This is the most developed of the new generation of reactors and its supporters claim that it might generate at between 2.2 and 3p per kWh, not as cheap as gas, but competitive with coal-fired plants (BNFL 2001). Interest has also been shown in an upgrade of the well-established CANDU heavy water moderated reactor, as developed in Canada and exported overseas, for example to India. The CANDU has the advantage that it does not need enriched uranium, although it still produces plutonium. This has led to concern about weapons proliferation issues.

A more radical option is the *high temperature reactor* (HTR), which uses graphite as a moderator and helium gas as a coolant. Several HTR systems have been tested over the years (including the Dragon reactor tested at Winfrith in the UK in the 1960/70s), although without too much success. However Eskom, the South African electric utility, is developing a small 100-MW HTR reactor, known as the *pebble bed modular reactor* (PBMR), so-called due to the sealed fuel pellets being encased in thousands of billiard ball-sized silica-covered spheres. These balls are contained in a hopper through which helium gas is passed. High thermal energy conversion efficiency (perhaps up to 40 per cent) is expected since the hot helium gas, exiting the core at 900 °C, actually drives the gas turbines – there is no secondary heat exchanger system. This could actually be a weakness of the design as, if some of the balls and the fuel pellets they contain rupture, the whole system could be contaminated with radioactive material, although it is claimed that the low-power density of the device (250 MW thermal output) would ensure that a fuel meltdown would be unlikely. Indeed, it is argued that, all being well, the spent fuel would remain safely encapsulated, ready for eventual disposal, without reprocessing.

In addition to the claimed safety aspects, it is argued that small modular reactors like the PMBR could be more easily mass-produced and, in theory, could be located nearer centres of energy demand, making them more suited to applications in the developing world. It has been estimated that the pebble bed modular reactor might have a capital cost of US$1,000 per kW installed and generate at between 2.4 and 4.3 cents per kWh but it will be some time before such claims can be tested.

The PBMR would still inevitably produce plutonium, as do all uranium-fuelled reactors. Given that plutonium is these days seen more as a problem than as a useful resource, there is still some interest in moving away from uranium entirely and devising reactors using another material – thorium. This is a weakly radioactive material which is actually more abundant than uranium. The disadvantage is that it cannot undergo fission directly, but mixed with a fissile material such as plutonium it can help sustain a chain reaction. As the result of the overall operating cycle, more plutonium is used up than is created, so the use of thorium–plutonium-fuelled reactors has been seen as one way to get rid of spare plutonium. Indeed, it is surprising that this idea has never really been taken up, unless, that is, plutonium production is the main attraction of reactors. However, it seems that it *is* still possible to extract a weapon-making material from thorium-based systems, so perhaps this option too is flawed. In addition a completely new fuel cycle infrastructure would have to be reconstructed – a huge task.

One of the most exotic new ideas is the so-called *energy amplifier*. Developed as a concept by Carlo Rubbia at the CERN Laboratory in Switzerland, this uses a proton accelerator firing a beam of neutrons into a sub-critical assembly of thorium coupled with plutonium, which can be used to transmute some long-lived radioactive isotopes into less dangerous forms. The energy output is seen as a by-product – the main aim is to convert some nuclear wastes into less hazardous forms. The system is seen as fail-safe, since the chain reaction can be instantly halted by switching off the accelerator.

Waste transmutation has obvious attractions, but a report to the European Parliament on new reactor options has noted that it was perhaps 'just nice physics' rather than an economically and technically practical option. It would be a slow process, and only some waste would be suited to transmutation. In addition to the energy amplifier transmutation plants, a complete power system designed on these principles would also require a complex and very expensive network of fuel processing and waste reprocessing plants if significant amounts of waste throughput were to be treated in this way. That would, the EP report says, only be possible, even in theory, for 'countries with a huge nuclear industry' (European Parliament 1999).

Finally, although it is also clearly a long-shot option, there is nuclear *fusion*. Rather than causing atoms of uranium or plutonium to split, as in nuclear fission, in a fusion reaction atoms of a form of hydrogen can be induced to fuse together at very high temperatures (tens of millions of degrees centigrade) to form helium, and this process is accompanied by the release of vast amounts of energy, much more than from fission. This is how the sun works. It is also the principle of the hydrogen bomb, in which the initial high temperature is created by exploding an atomic bomb, which then triggers a fusion reaction.

The process can in principle also be slowed down, and carried out under controlled conditions – for example by holding a cloud of very hot gas (called a 'plasma') in a vacuum away from containing walls by the use of magnets. To achieve fusion in these conditions very high temperatures are required, around 200 million °C – higher than in the sun. If the fusion reaction can be sustained a fusion reactor could, in theory, provide heat for power generation. Progress is being made with experimental devices like JET, the Joint European Torus at Culham in the UK, but, so far, no fusion reaction has been sustained for more than a very short time. A more advanced machine, the so-called International Thermonuclear Experimental Reactor, is planned, at a cost of £7.3 billion, in an attempt to get nearer to the conditions for a sustained fusion reaction. However, a full-scale power-producing reactor is generally seen as, at best, decades away.

If a full-scale power-producing fusion reactor could be built, then in principal, fusion would be better in resource terms than fission since, instead of uranium, it uses materials that are relatively abundant. The basic fuels in the most likely configuration to be adopted would be deuterium, an isotope of hydrogen, which is found in water, and tritium, another isotope of hydrogen, which can be manufactured from lithium. Water is plentiful but lithium reserves are not that extensive: even so it is claimed that they might provide sufficient tritium for perhaps 1,000 years, depending on the rate of use (Keen and Maple 1994).

Fusion reactions of the type likely to be used in reactors create large amounts of high-energy neutrons, which would have to be trapped, thus collecting energy from the reaction; that is how power would be generated. Although, unlike fission reactions, fusion does not produce direct wastes, other than helium gas, the containment materials and equipment in a fusion reactor would become very radioactive due to the high neutron bombardment. Since this could affect reactor performance and component functioning, these active materials will need to be removed periodically. These materials would only have 'half lives' of around 10 years, so that their radioactivity would decay much more rapidly than the radioactivity in some of the wastes generated in fission reactors, but these materials would still remain dangerous for perhaps 100 years (Keen and Maple 1994). There would still therefore, in effect, be a waste problem.

There could also be safety problems with the fusion reactor itself. Fusion reactions are difficult to sustain, so in any disturbance to normal operation the reaction would be likely to shut itself down very rapidly. But it is conceivable that some of the radioactive materials might escape, if for example the superhot high-energy 'plasma' beam accidentally came into contact with and punctured the reactor containment, before the fusion reaction died off. The main concern is the radioactive tritium that would be in the core of the reactor. Tritium, which is also used in nuclear weapons, is an isotope of hydrogen, and, if accidentally released, could be easily dispersed in the environment as tritiated water, with potentially disastrous effects. To put it simply, it could reach parts of the body that other isotopes could not.

Assuming the safety issues can be resolved, there is still a need to find a way of extracting useful energy out of a fusion reactor. At first sight this looks like an impossible task. On one hand there is a plasma at 200 million °C and on the other hand, assuming a conventional power engineering approach, we need to boil water to generate steam for a turbine. The mismatch seems vast: you can hardly put a hot-water pipe through the plasma. However, fortunately, it is not the heat of the plasma that would be tapped. Rather it is the energy in the intense neutron emissions that emerge from the fusion reaction that would be absorbed in some way and converted into heat; this in turn presumably being extracted by a conventional heat exchanger.

Given that it will be some time before a workable fusion reactor is available, not too much effort has been put into developing ideas for exactly how it might be used for power production. At some point in the future it may be possible to convert the energy emerging from the fusion reaction directly into electricity, but for the foreseeable future, if a fusion reactor can be built, perhaps rather strangely given its high-tech nature, it will still have to rely on a traditional steam-raising boiler to generate power.

Despite very large-scale funding over the years (some £20 billion has been spent worldwide so far) there is some way to go before fusion can be seen as anything more than a long-shot option. The physics has still to be fully resolved and a workable commercial device is at best decades away. It might therefore be considered to be irrelevant to current energy and environmental concerns. As has been indicated, there is a range of operational problems, and as yet few people would hazard a guess as to the economics of such systems. On this basis, there have been objections to the level of funding that has been given, and continues to be given, to fusion research.

It is also sometimes argued that fusion is irrelevant in longer term strategic terms, since, by the time a workable system has been developed, if it ever is, the world's energy problems should have been solved anyway. In which case, while it might be reasonable to continue some research as a long-term insurance, it could be argued that, given the inevitable scarcity of funds and given the urgency of our environmental problems, rather than trying to build a fusion reactor on earth, it might be more effective to make use of the one humanity already has – the sun.

Waste management and reprocessing

As can be seen, the nuclear industry is busy trying to find solutions to some of the key problems that its faces, so as to give itself a future. There are a range of new possibilities, even if most of them are relatively long term, and some, like fusion, very long term. However, there is a pressing short-term problem that also faces the industry – finding a way to deal with nuclear waste. Although some of the new reactor designs are claimed to be likely to produce less waste, they will still produce some.

Even without any new nuclear plants being planned, it remains a major problem. Over the next few decades, many of the first generation of reactors around the world will have to be decommissioned, and this will add to the problems of finding disposal routes for the radioactive materials. For example, as the UK Department for the Environment, Food and Rural Affairs has noted, in addition to the 10,000 tonnes of nuclear waste of various kinds already in existence in the UK, 'even if no new nuclear plants are built, and reprocessing of spent fuel ends when existing plants reach the end of their working lives, another 500,000 tonnes of waste will arise during their clean-up over the coming century' (DEFRA 2001).

At present, no permanent facility exists anywhere in the world for the long-term disposal of the most active high-level nuclear waste (see Box 6.2). The UK's waste management company, British Nuclear Fuels Ltd, claims that 'until one does, wastes can be stored safely and securely above ground for periods in excess of 50 years without the need for extensive store replacement and refurbishment'. However, this claim is not accepted by everyone, with, for example, the Nuclear Installations Inspectorate and the Health and Safety Executive regularly expressing concerns about safety. A particular concern has been the risk and impact of a terrorist attack. For example, whereas nuclear facilities have usually been designed to withstand accidental aircraft crashes, until September 2001, no one foresaw the risk of large passenger aircraft being deliberately used as weapons. The consequence of a successful attack could be very serious. A report to the European Parliament suggested that 'the long-term consequences of a release from the Sellafield high-level waste tanks could be much greater than the consequences of the Chernobyl accident, due to the large amounts of caesium-137 and other radioisotopes in the Sellafield tanks' (STOA 2001).

An allied issue is the question of what to do with the plutonium that has been extracted from the spent fuel by reprocessing. So far it has not been seen as a 'waste' but as a valuable material either for bombs, or for fuelling fast breeder reactors to generate yet more plutonium. However, as we have seen, there are no longer any major fast breeder reactor programmes, and the end of the Cold War means that there is less need for weapons material. In any case, there are alternative techniques to reprocessing for obtaining military plutonium.

As noted earlier, the industry's favoured option for dealing with plutonium at present is to mix plutonium oxide with uranium oxide to

Box 6.2

Nuclear waste disposal

'It would be morally wrong to commit future generations to the consequences of fission power on a massive scale unless it has been demonstrated beyond reasonable doubt that at least one method exists for the safe isolation of these wastes for the indefinite future' (Royal Commission on Environmental Pollution 1976).

More than twenty-five years on, the UK is still not much closer to finding a solution to the nuclear waste problem. The main difficulty has been in finding acceptable sites for waste repositories. An attempt by the UK Atomic Energy Authority to find suitable sites for disposing of vitrified high-level waste in Scotland, Cornwall, Wales and Northumberland was abandoned in 1981, following local opposition. An attempt to assess the suitability of sites at Billingham on Teeside, and Elstow, near Bedford, for low- and intermediate-level waste repositories, was abandoned in 1987, after extensive protests by people in the selected areas (Blowers and Lowry 1991).

In 1997, the proposal by the UK nuclear waste company Nirex to build a test site for an underground repository for high-level wastes at Sellafield also failed to obtain government approval. The then Secretary of State for the Environment said he was 'concerned about the scientific uncertainties and technical deficiencies in the proposals presented by Nirex [and] about the process of site selection and the broader issue of the scope and adequacy of the environmental statement'.

The UK Labour Government, elected shortly after this decision, was therefore confronted with the need for a new technical approach and policy on nuclear waste management. It promised a wide-ranging consultation, which was launched in 2001. However, the consultation and review exercise is not scheduled to be completed until 2006, so it will be some time before the issue can be resolved, and even longer before a waste site could be operational.

A similar situation exists in France – a decision on where to locate a final repository for its large stocks of waste will not be taken until 2006, and any such repository would not be ready to receive waste until around 2020.

By contrast, the USA is trying to push the pace on faster. Pursuant to his commitment to resuscitating nuclear power as an energy option, in May 2001, Vice President Dick Cheney indicated support for using Yucca mountain, in a remote part of Nevada, as a site for the USA's wastes. This site has actually been under investigation for nearly twenty years, but there are still many political and regulatory hurdles to negotiate before it can be guaranteed as a viable resting place for the USA's high-level nuclear wastes.

Sweden has a repository in operation for the less active low- and medium-level nuclear waste, and an interim store for spent fuel, having adopted what some call a 'Rolls Royce', no-expense-spared approach following widespread and sophisticated public consultation, linked to the plan to phase out the use of nuclear power. No long-term repository for the most dangerous high-level wastes exists as yet. Possible sites for such a deep repository for spent fuel are being investigated, but a decision on location has yet to be made and such a repository could not be in operation until the mid-2010s.

Finland used to have its spent fuel reprocessed in Russia, but now plans to encapsulate and store it all underground in bedrock. Like Sweden, it has carried out extensive public consultation on this issue. The construction of its final disposal facility is scheduled to start in the 2010s and the facility should be operational after 2020.

Even if there is no more expansion of nuclear power around the world, with temporary stores, usually on site at nuclear plants or at reprocessing plants like Sellafield, beginning to fill up, clearly this is an issue that the world's nuclear power users will have to face. Interestingly, Russia has offered its services as a repository for other countries' nuclear wastes, but, so far, no one has taken up this option.

produce mixed oxide fuel (MOX) for conventional reactors. However, not everyone is convinced that MOX is a good or viable option, not least since MOX is much more expensive than conventional uranium fuel, by perhaps a factor of five (Garvin 1998), and using it in reactors does inevitably generate more plutonium, from the uranium-238 part. Critics also claim that, far from providing an easy way to deal with plutonium stocks, only plutonium that has been reprocessed relatively recently can be used for MOX. Plutonium that has been stored for longer than around four or five years, including, presumably, plutonium from redundant nuclear weapons, cannot be used for MOX unless it is specially treated, due to the build up of contamination from radioactive decay products, thus adding further to the cost (OECD and NEA 1989).

It is also argued that there are other ways to immobilize and secure plutonium so as to keep it safe from misuse, for example by mixing it with nuclear wastes (or of course leaving it in un-reprocessed spent fuel). Moreover, although it would generate energy from what would otherwise be wasted material, not only is this expensive, it is risky moving it around to reactors, since it is possible to use it for weapons production (it is easier to get at the plutonium in unused MOX fuel than in the fiercely radioactive spent fuel rods). Consequently, shipments have to be heavily guarded. Moreover, as with any other nuclear fuel, once used in a reactor, the highly active spent MOX fuel would have to be dealt with. But the crucial point is that the use of MOX fuel will only make a small dent in the plutonium stockpile, since this is growing all the time due to the continuance of reprocessing operations.

As the long history of accidental leaks at Sellafield (and at its French equivalent in Brittany) has illustrated, reprocessing is arguably the main weak point in the nuclear fuel system, whatever type of reactor is used. The USA does not reprocess its spent fuel and only a minority of EU countries now favour reprocessing – dry storage of used fuel rods is becoming the favoured option. The main reason for reprocessing was to get access to the plutonium in spent fuel, but this process creates a lot of low- and intermediate-level wastes. This process also has major implications for radiation exposure both to nuclear workers and to the general public – it involves a series of chemical separation stages which can increase the risk of minor accidents and leaks and involves more handling of radioactive material of various grades by workers. It has been estimated that around 79 per cent of the collective radiation dose associated with the complete nuclear fuel cycle comes from reprocessing activities. By contrast, electricity generation in nuclear plants accounts for only around 17 per cent of the collective dose, uranium mining and milling only about 2.2 per cent and waste disposal only just over 1 per cent – all measured on the basis of the 'dose per kWh of power finally generated' (European Parliament 1999).

On this basis, it is clear why some people see reprocessing as a problem, and why strenuous efforts have been made to get tougher regulations on the levels of allowed emissions from reprocessing plants. For example, the current agreement is that liquid discharges should be reduced 'to near zero' by 2020. However, some opponents do not believe that even this target, which will cost the industry a lot to meet, will be sufficient, since there is still the risk of accidents. They would prefer reprocessing to be abandoned altogether.

The counter-argument is that, since reprocessing separates out the plutonium and uranium, it leaves a smaller amount of high-level nuclear wastes to deal with. While that may be true, the chemical separation process and associated handling activities lead to large amounts of low- and intermediate-level wastes, and, overall, it increases the potential level of radiation exposure to workers and the public. It would be far safer, it is argued, to leave the spent fuel un-reprocessed and store it. That would have the added bonus that the plutonium would then be much more secure from theft. Moreover, this would also be likely to be the cheaper option. Certainly, reprocessing is expensive, and it would seem unwise to incur the significant costs and risks, unless you want to get access to the plutonium, so as, for example, to produce MOX.

Finally, there is the issue of plant *decommissioning*. As reactors come to the end of their useful lives or are phased out, they have to be dismantled and the active materials they contain made safe. The UK industry's preference is to remove the non-active parts of the plant, but leave the reactor core *in situ*, sealed with weatherproof cladding in a so-called 'safe store', for 135 years or so. It would then be less active and therefore safer and cheaper to demolish. Postponing full decommissioning is not popular with environmentalists, who argue that this amounts to leaving it to future generations to deal with. The UK Nuclear Installation Inspectorate have suggested that, in fact, the safety case may only imply a deferral for around 40 to 50 years, at least for some reactors (NII 2001). That is evidently the sort of timescale most other countries are considering. However, a longer delay has the economic attraction that, if relatively small amounts of money are put aside from current nuclear earnings, then, given that this money will be invested over a long time period, sufficient capital should be available to deal with final decommissioning. For example, if decommissioning the reactor cores can be delayed, for say 135 years, it has been estimated that the discounted liability for decommissioning the Magnox and advanced gas-cooled reactors would reduce by between 45 per cent and 50 per cent respectively (NAO 1993). With BNFL's total historic liabilities for plant decommissioning and site clean-up (including the Sellafield site) having been put at up to £34 billion, the attraction of discounting costs into the future becomes clear – it can produce significant changes in the financial burden.

A nuclear revival?

Assuming the waste and safety issues can be resolved, a further crucial prerequisite for a revival of nuclear power would be improvement in the economics of nuclear operations and the main hope for the future is, therefore, new technology. Some of the new nuclear plants currently being developed, as described earlier, are claimed to be likely to generate at somewhat more competitive prices.

For the moment however, these figures are just speculation, and, in order for new nuclear plants to achieve viability, most analysts of nuclear economics also have to assume some form of subsidy, or increases in the cost of rival technologies.

For example, at present, Finland is the only Western European country currently planning nuclear expansion, having concluded that new nuclear plants, including building, running and decommissioning costs, would be competitive in terms of price with all other alternatives, including combined-cycle gas turbines. However, this analysis was based on the assumption of a 50 per cent rise in gas prices. This price rise may well happen, but then many other technologies would also be competitive.

In its submission to a review of energy security carried out by the House of Commons Trade and Industry Select Committee in 2001, the British Nuclear Industry Forum noted that, even given the new technology that might be available within ten years, there was still a gap between the wholesale electricity price of around 1.8p per kWh and the cost of new nuclear-generated electricity of about 2.5p per kWh. What was therefore required, they argued, was some sort of financial assistance to bridge that gap, preferably one that would provide a subsidy of about 1p per kWh (Select Committee 2002).

In terms of subsidies, the nuclear industry around the world has done well over the years. Indeed, it could be argued that it would not exist without subsidies and other forms of financial aid, some of which continue to protect it from competition. For example, in 2002, the UK government proposed that around £48 billion in historical liabilities (e.g. for waste clean-up and plant decommissioning) that had been built up by the state-owned company BNFL and the Atomic Energy Authority, should be transferred to a new public body, thus presumably paving the way for the subsequent partial privatisation of BNFL, possibly via a public–private partnership arrangement.

The continuing need for subsidies was highlighted in September 2001, when British Energy, the UK's main nuclear plant operator, which had been set up as a private company in 1996, sought financial support from the government, since it could no longer survive in the increasingly competitive UK electricity market. It was reported to be losing £4 on each MWh sold. £650 million was provided as an interim loan while the government looked for ways to resolve the problem. Some form of continuing public support seems to be inevitable if the company is to continue operating its power plants.

The nuclear industry also benefits from legislation exempting nuclear companies from unlimited liability for accidents, a cover currently provided in the USA by the Price-Anderson Act and by similar arrangements in the EU. For example, in the UK, under the Nuclear Installations Act (1965), the operator's liability was limited to a maximum

£140 million per accident and, if damage caused exceeded that amount, public funds would be made for the payment of compensation to a total amount of £300 million. However, in the wake of the terrorist attacks in the USA in September 2001, the UK insurers withdrew cover from nuclear facilities, and the government had to step in and take over the full liability.

France is often held up as a example of a country where nuclear power is economic, and certainly the fact that they launched a major programme based on serial production of standard reactor designs had the benefit of economies of scale. However, this state-led construction programme was funded not by the charges made to consumers for their electricity, as in the UK and some other countries, but at least in part by borrowing on the international money markets. This led to low consumer prices but it also left France with a large outstanding debt to service. Unfortunately, demand did not rise as much as expected when the programme was planned in the 1960s, and France had to sell off its excess nuclear electricity at relatively low rates to other European countries, including the UK. But given the low prices charged for this electricity, the interest on the initial capital borrowed still represents a sizeable proportion of the income generated, as the data in Box 6.3 illustrates. Indeed, it seems initially that the debt repayments were greater than the income raised.

Box 6.3

French nuclear debt

According to the World Nuclear Association, France's nuclear power programme cost some FF 400 billion in 1993 currency, excluding interest during construction. Half of this was self-financed by Electricité de France, 8 per cent (FF 32 billion) was invested by the state but discounted in 1981, and 42 per cent (FF 168 billion) was financed by commercial loans. In 1988 medium- and long-term debt amounted to FF 233 billion, or 1.8 times EdF's sales revenue. However, by the end of 1998, EdF had reduced this to FF 122 billion, about two-thirds of sales revenue (FF 185 billion) and less than three times annual cash flow. Net interest charges had dropped to FF 7.7 billion (4.16 per cent of sales) by 1998.

Source: From the WNA's evidence to the PIU Energy Review: http://www.world-nuclear.org/wgs/wnasubs/energyreview/index.htm

Following the election of George W. Bush in 2000, the USA began to reconsider the nuclear option and in 2002, the Bush administration allocated $3 million to streamline applications for building new plants – which would involve the first new orders since 1978. In addition, the US government plans to match utility investment up to $48.5 million to license new sites. Even in the UK, where the government ostensibly has a policy of 'diminished reliance on nuclear power', state support for nuclear R & D has continued. For example, nuclear R & D (including for fusion) was allocated £24 million out of the total UK budget for energy R & D of £50.7 million in 2000–1 and the projections for 2003–4 are that the nuclear share will rise to £52.3 million out of a total energy R & D budget of £106.1 million. Some of this nuclear R & D is for work on decommissioning and clean-up operations (including £32.2 million in 2003–4 in support of the Russian nuclear programme), but, even so, nuclear R & D is still attracting around half the total energy R & D budget, with R & D on renewables trailing behind at £13.6 million in 2000–1 and £23 million in 2003–4 – less than half the nuclear allocation.

However, as the economic problems facing British Energy illustrated, despite these various subsidies, nuclear power still does not seem to be viable in the UK. An independent study has suggested that, even with a significant economic boost from some form of carbon tax or carbon credit, it would still be hard for nuclear to be competitive with gas in the UK's liberalised electricity market (Pena-Torres and Pearson 2000). For example, the Climate Change Levy, which was introduced in the UK in 2001, imposes a 0.43p per kWh surcharge on the electricity used by most companies. However, if electricity from nuclear power plants was to be exempted from this levy, as some have proposed, the resultant price reduction would still not be sufficient to make new nuclear power plants competitive with gas plants. As noted earlier, the nuclear industry has suggested that it would need a subsidy of around 1p per kWh to bridge the gap.

Nuclear power and the developing world

The situation elsewhere might be different, for example, in the developing world. The nuclear industry was very keen to be allowed to obtain support for new projects overseas from the Clean Development Mechanism (CDM), the emission credit scheme first proposed at the UN Climate

Change conference held in Kyoto in Japan in 1997. The CDM is meant to support projects in developing countries which avoid greenhouse gas emissions, and award the developers 'credits' for so doing. Given that these credits can be traded, in effect the cost of the project is reduced. It has been estimated that this might reduce the cost of nuclear plants by up to 20 or 30 per cent. However at the reconvened Conference of Parties to the Kyoto agreement held in Bonn in 2001, it was decided, after pressure from the EU, that nuclear projects should not be eligible for CDM credits, with opponents to nuclear inclusion arguing that it was not a clean, safe or sustainable option, nor a useful tool for economic development (NIRS 2000).

While there are some prospects for a nuclear revival in Western countries, this does not seem likely to be on a large scale, and it is perhaps therefore not surprising that, even without the CDM, the main emphasis for Western nuclear companies currently seems to be export orders, with the developing world being an obvious target.

This raises the issue of the role of nuclear power, in economic development. Some developing countries evidently see nuclear power as part of the industrialisation process. 'High' technology developments like this may benefit some members of the technical and economic elites in some developing countries, but it can also be argued that importing capital-intensive nuclear power technology does not seem to be the best bet for those Third World countries that are struggling with already large foreign debts, or those whose populations in the main need cheap, simple, locally accessible sources of power. There are also worries that part of the attraction of 'going nuclear' is that it can provide a means of developing nuclear weapons. There are other routes to obtaining the necessary nuclear materials, but civil nuclear programmes provide one potential source.

Although most, but not all, countries around the world have signed nuclear non-proliferation agreements, nuclear expansion raises concerns about the problem of illegal diversion of materials for making nuclear weapons like plutonium. These materials are carefully controlled, but a black market has grown up, particularly since the collapse of the Soviet regime. Adding to the security problem by building more reactors would seem to be unwise.

The ex-Soviet nuclear programme, which provides around 12 per cent of the electricity used in the region, also presents other problems: the safety of some of the reactors worries many observers, but few of the new

Central and Eastern European states can afford to clean them up – or to shut them down, since they need the power. Funds to help improve safety have been made available from the EU and elsewhere, and also for the construction of some new reactors. Given that Russia has one of the largest reserves of gas in the world, this may appear a little strange, but then Europe is relying on having access to these gas reserves when North Sea gas becomes scarce, and Russia needs the foreign exchange it can earn from selling its gas to the West. In this context it is perhaps less surprising that they and the West are keen to keep Russia's nuclear plants going and support the construction of new ones.

A nuclear future?

Clearly, the nuclear industry is keen for the use of nuclear technology to increase, and sees climate change concerns as its possible saviour. At the very least, the industry does not want the nuclear contribution to diminish. Thus, speaking at a World Nuclear Association Conference on 6 September 2001, BNFL's Chief Executive Norman Askew argued that 'Nuclear energy must continue to play a significant role in the UK's base-load electricity generation. Without nuclear's contribution this country cannot have a continued secure, diverse and environmentally-friendly energy supply.' But equally, the environmental opponents of nuclear power are keen to see it contract – and argue that it is foolish to try to resolve one problem (climate change) by introducing another (pollution by nuclear radiation). They believe the money would be better spent on renewables and energy conservation. Thus the US-based Nuclear Information and Resource Center has argued that nuclear expansion would be a 'lose–lose' option for the environment:

> Not only will there be an expanded nuclear industry, with increased production of radioactive waste and the constant risk of catastrophic accidents, but every dollar spent on nuclear power will be diverted from the development of sustainable energy systems and effective measures to combat climate change.
>
> (NIRS 2000)

There could of course be political breakthroughs, particularly in the USA, where, in 2001, the Bush administration indicated a desire to rethink its policies on nuclear power. In particular, US Vice President Dick Cheney has backed nuclear expansion, seeing nuclear power as a 'safe, clean and very plentiful energy source'.

However, in the UK, the comprehensive review of energy policy over the next fifty years, carried out in 2001 by the Cabinet Office's Performance and Innovation Unit (PIU), saw nuclear power as likely to remain relatively expensive. Even with the new reactors that might be ready for commercial use within fifteen to twenty years, generation costs for new plant were put at between 3 and 4p per kWh. Consequently, the PIU, in effect, relegated nuclear power to a longer term 'insurance' role, to be called on in case their preferred options, renewables, combined heat and power and energy efficiency, failed to deliver. The PIU concluded that 'there is no current case for public support for the existing generation of nuclear technology', although it added 'there are however good grounds for taking a positive stance to keeping the nuclear option open' (PIU 2002).

This 'leave the nuclear option open' policy was clearly also supported by the major *World Energy Assessment* carried out by the UN Development Programme, the UN Department of Economic and Social Affairs and the World Energy Council. It concluded that 'if the energy innovation effort in the near term emphasizes improved energy efficiency, renewables, and the decarbonised fossil energy strategies, the world community should know by 2020 or before much better than now if nuclear power will be needed on a large scale to meet sustainable energy goals' (WEA 2000: 318).

Summary points

- Nuclear power plants do not generate greenhouse gases directly and so might play a role in responding to climate change.

- Although there is currently a glut of uranium, reserves might become scarce and expensive if an attempt were made to respond to climate change by building large numbers of conventional nuclear reactors.

- Fast breeder reactors could extend uranium reserves, but at unknown cost, and the waste and plutonium proliferation problem would be increased.

- There are some new reactor technologies emerging that might be safer and cheaper – but it is too soon to know for certain whether the nuclear industry can resolve all the problems facing nuclear power.

- Nuclear fusion could, if successfully demonstrated, offer a new energy source with a reasonably long life, but it is a technological 'long shot' with its own safety and economic problems.

- The nuclear waste issue remains unresolved, and there are growing fears about the risks of proliferation and theft of weapons-making material and about terrorist attacks on nuclear facilities.

- The debate on nuclear power continues, with opposition still being strong, but with one view being that we should wait to see whether we need to revive this technology, focusing meanwhile on the renewables and energy conservation.

Further reading

In addition to the suggested readings at the end of the last chapter, for a useful overview of the new nuclear technologies that are emerging, see Chapter 8 of the UN/World Energy Council's *World Energy Assessment*, produced in 2000. It can be freely downloaded from http://www.undp.org/seed/eap/activities/wea/index.html.

The Royal Society/Royal Academy of Engineering report 'Nuclear Energy: the future climate', published in 1999, also contains some useful coverage of the new technologies and the issues implied by their development. It can be accessed from http://www.royalsoc.ac.uk/policy/reports.htm.

In February 2003, the UK government produced a White Paper on energy, 'Our energy future: creating a low carbon economy', which adopted the PIU's recommendation that nuclear power should be left as an 'insurance' option, for the longer term. See http://www.dti.gov.uk/energy/whitepaper/index.shtml.

7 Renewable energy

- Solar power
- Wind power
- Wave power
- Tidal power
- Wastes and energy crops
- Hydroelectricity
- Geothermal energy

Natural energy flows and sources such as sunlight, the winds, waves and tides, offer a relatively clean, safe and above all sustainable source of power. This chapter explores the renewable energy options, looking at the basic technology and at some current developments, and then reviews the overall global resource potential of renewable energy. In the following chapter we review renewable energy developments around the world.

Renewable energy technology

Renewable energy is so called because it relies on natural energy flows and sources in the environment, which, since they are continuously replenished, will never be exhausted. So they meet our first criteria for sustainability. In what follows, to provide a context, we first look at the basic technology before moving on in the next chapter to look at developments around the world.

The basic technology is very varied, reflecting the variety of natural energy sources. As has been indicated, most sources of renewable energy are the result, directly or indirectly, of the impact of solar radiation on the planetary ecosystem – the exceptions being tidal energy flows and geothermal heat. Incoming solar radiation provides energy for plant and animal life and drives the hydrological weather cycle – evaporating sea water and creating rain, which feeds into rivers and streams. Local differential heating of the atmosphere, sea and land also causes winds, which move over the seas creating waves. By contrast, tidal energy could be thought of as lunar power, given that the main component is due to the effect of the gravitational pull of the moon on the seas, although the sun's gravitational pull also plays a part. Geothermal energy is the result of residual heat deep in the planet, topped up by radioactive decay processes.

Renewable energy is actually already in widespread use: around 20 per cent of the world's electricity already comes from conventional hydroelectric dams, and in many countries wood and animal dung provide the only sources of power for cooking and heating.

Now, however, a whole new range of renewable energy technologies is emerging. The sections that follow explore some of the key new technologies for harvesting renewable energy, looking first at solar power, including photovoltaic solar cells. We then move on to look at the basic physics and technology associated with the use of wind, wave and tidal energy, and then look at energy crops, focusing on short rotation coppicing.

Solar power

The sun provides the basis for life on earth and delivers sufficient energy to each square foot to meet all our needs – if that energy can be tapped efficiently. In the past, human beings have tried to do that via agriculture, using wood as a fuel and by using the indirect solar energy represented by winds and streams. More recently use has been made of the stored solar energy of fossil fuels – coal, oil and gas.

However, the sun's heat can also be used more directly. Around the world in recent years there have been large-scale experiments with solar power; for example, via giant solar heat-concentrating mirrors, parabolic troughs and dishes, tracking the sun across the sky and focusing its rays so as to raise steam for electricity generation. Large-scale 'solar thermal' plants like this are becoming increasingly popular in desert areas of the USA and elsewhere. Another idea, currently being looked at in Australia, is the so-called 'solar tower'. Hot air from a very large solar-heated conservatory around the base of a giant 1-km tall chimney, would rise by convection and turn air turbines within the tower. A 50-kW plant was tested in Manzanares, Spain, in the 1980s, but the system proposed in Australia is much larger – 200 MW.

Solar energy can also be utilised on a smaller scale, via roof-top solar collectors (see Figure 7.1), which are in widespread use in many countries around the world, for example in the Mediterranean region. Even in the UK, flat-plate solar heat collectors, looking something like radiators mounted on roof tops and plugged into a hot water system, can typically cut average yearly domestic fuel bills by a half. However, in most locations in the UK, the overall economics of solar space and water heating is currently not that attractive compared with cheap gas heating: commercial solar energy systems have payback times of five to ten years or more.

An alternative is the more cost effective *passive solar* concept. Conventional roof-top solar collectors use small pumps to drive the heated water around the heat circuit, but you can also collect useful amounts of heat if you have large south-facing glazed areas. It is much like the greenhouse concept – and involves no moving parts. Hence the term 'passive' as opposed to conventional 'active' solar collectors, with pumps.

Typically, with a well-insulated house, overall annual fuel bills can be cut by a third in this way. Large-scale solar atriums for offices and

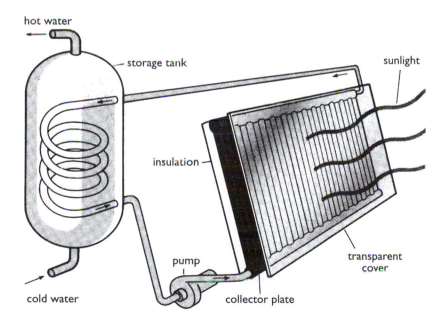

Figure 7.1 *Flat-plate solar collector for space and/or water heating*
(*Source:* Open University T362)

commercial buildings are now a familiar sight, and, at the more lowly level, many people have added solar heat-trapping conservatories to their homes. In addition, in some places in the world, it makes sense to store solar heat from the summer to use in the winter, by heating up thermal masses with solar-heated air or water, and then, in winter, passing air or water over to be warmed. This is expensive and cannot be easily done on a small scale. However, there are economies of scale. A large interseasonal heat store has been constructed in Sweden consisting of crushed rocks buried in a well-insulated cavity underground, feeding heat obtained in summer from a large array of solar collectors to a communal district heating network, which supplies 550 dwellings with winter heat. Shorter heat storage options include the use of solar ponds – an idea being explored in Israel.

However, the big breakthrough is likely to be in the *photovoltaic* solar field. Photo cells, like those on cameras and pocket calculators, convert sunlight directly into electricity. The only problem is that they are expensive. They were initially used mainly for powering space satellites, but developments in the semiconductor field have gradually brought prices down. Within a decade or so they are likely to be competitive with conventional power sources.

In this case, photovoltaic (PV) cells are likely to be used widely – even for domestic supplies. They are already in use in some outlying rural areas where there is no grid electricity and the alternative is diesel, or nothing – in the Australian outback, in desert areas and in African villages – for water pumping, running refrigerators for key medical supplies or powering remote telecommunications equipment.

Soon there may be much wider scale use, especially since pv cells can be substituted for roof or fascia cladding on buildings, thus saving on the cost of conventional roofing or wall materials, and offsetting the cost of the PV cells. Many projects around the world have already adopted this idea, particularly in Germany, which is expected to have 14,000 PV systems installed by 2005. Clearly, office buildings are an attractive option in this context, given their daytime occupancy, but domestic projects have also emerged for example in Denmark. The UK has made a slow start in this field, despite some pioneering efforts. See Box 7.1. However, in 2001 the government announced a £20 million funding programme based on grants to meet 50 per cent of installation costs, which could lead to the installation of over 3,000 PV systems. At the same time costs are dropping as new types of cell materials have been developed that are more efficient – commercial modules can be up to 15 per cent efficient and some laboratory devices have efficiencies of 25 per cent or more, so that the need for subsidies should lessen. In terms of costs, ease of manufacture is an obvious issue and thin-film amorphous silicon technology has proven to be attractive, despite its lower efficiency. There are also exciting prospects for some of the newer polymer cells.

Box 7.1

The Oxford solar house

The UK may have been slow to take up solar photovoltaics, with only around 1.5 MW installed by 2002, compared with nearly 2 GW globally, but there are, nevertheless, some interesting projects around. One of the first domestic housing projects was the Oxford Ecohouse built for Sue Roaf, a professor at the Oxford School of Architecture. It included a large 4 kW peak array of PV cells on the roof together with a 5 m^2 solar heat collector, passive solar features and good insulation, coupled with a wood fuel back-up stove, and a highly efficient gas-condensing boiler for space and water heating top-up, when needed. It was finished in April 1995 and in its first year generated 2,937 kWh from the PV cells. Of this power, 57 per cent was exported via the standard link to the national grid, but the local electricity company only paid 2.8p per kWh for this power. Although the house consumed only 2,689 kWh over the year, and thus was the net generator of the power, not all this power was available at the time needed, and so occasionally it had to import some power to top up. This cost 7p per kWh, so there was a net bill of £60. But it would otherwise, without the PV, have been around £190, so, after deducting metering rental, the house still generated a net income of around £100 p.a.

Since then some local electricity companies in the UK have developed so-called 'net metering' arrangements for PV solar, which charge/pay an agreed amount for the net flow of power, whichever way it flows, typically around 5p per kWh. Under such schemes, the house would make a significant profit, despite the fact that Professor Roaf also runs a small electric vehicle from the power the roof produces. Of course, given that the total system and installation cost for the Oxford Ecohouse was £25,000, it will take a while to pay back at that rate, but even so it does show what can be done. Certainly, it is interesting having a roof that earns its keep.

See S. Roaf, M. Fuentes and S. Thomas, *Ecohouse: a Design Guide* (Architectural Press, 2001).

'Solar electric: building homes with solar power' can be obtained from Greenpeace, Canonbury Villas, London N1 2PN. Tel: 020 7865 8100.

Given these trends, the use of pv seems bound to spread. Certainly, there are very large PV deployment programmes in the USA, Japan and Germany – with, for example Germany having a 100,000 roof programme and Japan aiming for 70,000. Overall, solar pv is increasingly seen as being likely to be a major energy source in the decades ahead.

Environmentally, there would seem to be few problems, at least in terms of the use of pv cells, although there are some question marks associated with their manufacture. This is very much a high-tech industry, using exotic and often hazardous chemicals – potentially representing a significant health and safety problem for workers. There could also be ex-plant pollution issues to contend with. However, assuming these potential problems can be avoided, pv could become a key new renewable resource.

Wind power

In contrast to PV, wind power is already a significant energy source. The winds are an indirect form of solar power, and they have been used for centuries as a source of energy. More recently wind power has become one of the more successful renewable energy technologies. The USA took the lead, with the result that California's wind farms supply the energy equivalent of the domestic electricity requirement of a city like San Francisco. Europe then took over, with Germany taking the lead, and wind projects now exist in many other parts of the world. By 2002, the total generating capacity around the world was around 24,000 MW, with costs dropping dramatically as the technology developed. For example, by 2002, wind projects were going ahead in some parts of the UK that were competitive with gas-fired plants.

The implications of the basic physics of wind turbines can be relatively easily appreciated. We will focus on propeller-type devices, which have proved to be the most popular, although vertical axis designs may have a role in some circumstances (they have the advantage of accepting wind from any direction and of having the generator unit on the ground rather at the top of the supporting tower). The energy collected from a wind turbine is proportional to the area in the circle swept by the propeller blade, i.e. πr^2 where r is the blade length. The result of this 'square law' is that, for example, doubling the size of the wind turbine quadruples the power output, meaning that large turbines are much more effective. But beyond a certain size the blades face stress (and fabrication cost) limits and currently the largest size commonly used is between 1 and 2 MW rated capacity (i.e. at full power) with blade diameters of between 60 and 65 metres. However, larger devices are emerging, of 3, 4 and even 5 MW, mostly for use offshore.

The power in the wind is proportional to V^3, the cube of V, the wind speed. (The kinetic energy is $\frac{1}{2}mV^2$, where m is the mass of the air intercepted, which in turn is proportionate to V times the area swept.) So even a slightly higher wind speed site pays dramatic dividends in terms of available power, even more than from increasing the blade size.

However, the actual amount of power wind turbines can produce is less than the power in the wind flow they intercept. In part, this is a matter of design and operational factors, but fundamental aerodynamic losses mean that the maximum amount of power that can theoretically be extracted is just under 60 per cent of that in the flow and there are also inevitably mechanical and electrical losses in conversion.

Wind turbines are often grouped together in 'wind farms', so that connections to the power grid can be shared, as well as control systems and road access for maintenance. Typically, a separation of between 5 and 15 blade diameters is needed between individual wind turbines, to prevent turbulent interactions in wind farm arrays. This means that wind farms can take up quite a lot of space, even though the machines themselves only take up a small fraction of it, and this has led to some objections. It is argued that there would be insufficient room in countries like the UK to generate significant amounts of power. As Box 7.2 shows, it is relatively easy to work out the rough land area requirements for a wind farm, and then to calculate how much power might be generated in a country like the UK.

As Box 7.2 shows, even given the intermittancy of the wind, it would seem possible to replace 10 per cent of the UK's conventional power generation capacity with 2-MW wind turbines by using around 1 per cent of the UK's total land area, depending on the assumptions used. Obviously, this is only a very rough estimate, based on some broad assumptions. The actual power output in practice would depend on the sites, and the operational patterns, and, in reality, more power might be obtained from wind turbines covering less area. Using larger machines or using array separations of less than 10 diameters would have the same effect. Equally, however, since wind turbines are visually intrusive, there will be specific siting constraints which could reduce net power availability: some high wind speed sites may not be acceptable. This issue

Box 7.2

Power produced and land used by wind farms – a rough calculation

We can calculate the power output and area covered by a typical wind farm by making some very broad assumptions, using round figures, as follows. A 10 by 10 array of 100 2-MW machines (200 MW in all) each with say 60-metre diameter blades, and with 10-diameter separations, would cover an area of 6 km by 6 km. However, only about 1 per cent of this area (i.e. 360 m^2) would actually be occupied by the bases of the turbine towers; the rest of the land around the wind-turbine bases could be used for agricultural purposes.

Of course these wind turbines would not be able to operate continuously at full power since the wind is intermittent. Typically, given the variability of the wind, wind turbines in the UK can only deliver power on average for about 30 per cent of the time: this figure is known as the 'load factor'. To make a fair comparison with conventional plants, it is important to realise that, although fossil- or nuclear-fuelled plants do not have intermittent energy inputs and therefore have much higher load factors, they typically can still only achieve load factors of around 70 per cent. Consequently, to generate the same amount of power from each, on average you would need just over twice as much wind-farm generating capacity as you would conventional capacity. Put the other way around, you would expect wind turbines on average to generate less than half as much continuous power as conventional plants with the same capacity – specifically, 43 per cent as much (30/70 × 100 per cent). In the UK the government gives wind turbines and other renewable energy devices using intermittent sources a rating in terms of their **declared net capacity** (DNC) to reflect this comparison, with the conversion factor for wind being defined as 0.43 of the installed capacity. So the equivalent DNC generation capacity of a wind turbine is considered to be 43 per cent of its full rated capacity.

On this basis, although our hypothetical 100 turbine wind farm would in principle have a generating capacity of 200 MW, in practice it would be equivalent to only 86 MW DNC. The UK currently has 60 GW or so of conventional installed capacity, leaving aside existing renewable sources like hydro and the combined heat and power element, so if 10 per cent of this were to be replaced by wind turbines we would require around 70 wind farms of this size. The total tower base area covered would be around 25 km^2, while the total area covered by the complete wind-farm arrays would be 2,520 km^2. The latter is around 1 per cent of the UK's total land area (250,000 km^2).

So, it would seem possible, even in a crowded country like the UK, to obtain around 10 per cent of the country's current electricity requirements by using, say, 1 per cent of the land area. That, as it happens, is roughly the area that has been estimated as likely to be suitable and available for wind-farm projects in the UK without significant intrusion (Clarke 1988).

Note that in the analysis above the DNC figures in effect are based on using *average* power figures for wind – averaging out the variations over time. We will be looking at the issue of intermittency of renewable energy sources in Chapter 9. Suffice it to say here that, although the *operational* value of the power actually delivered will of course be determined in part by when it is available in relation to energy demand, in practice grid systems can balance out quite large variations in input. Moreover, the main *environmental* value of wind energy, or any other renewable, is a reflection of the fact that every kWh generated will replace a kWh of energy generated by conventional power plants (Milborrow 2001).

will be explored in detail later, but for the moment it is perhaps worth noting that, by 2002, there were around 1,000 wind turbines installed in the UK, although the British Wind Energy Association has indicated that it envisages no more than around 2,500 new wind turbines being installed on land in the UK by 2010, so that in total wind would be delivering very roughly about 5 per cent of average UK electricity requirements.

The offshore potential is much less constrained. For example, between 100 and 150 TWh p.a. could in theory be obtained from sites in shallow water off East Anglia, i.e. over a third of the UK electricity requirements, and more from sites further out. Indeed in theory offshore wind could supply all the electricity the UK needs.

The UK is particularly blessed with good sites, but other countries also have some good locations, and have moved quite rapidly to develop them. By 2002, there was was nearly 100 MW of offshore capacity in place in the EU. The Netherlands, Denmark and Sweden have been pioneers, and they are now being followed by Germany, Eire, Belgium, France and the UK. Germany has a very ambitious programme, with a target offshore wind capacity of 10 GW by 2030, supplying around 25 per cent of the country's electricity. Overall the EU's total ultimate offshore wind resource has been put at around 900 TWh p.a.

The UK started its offshore wind programme relatively late, but has plans for eighteen sites around the coast. Although there are extra costs associated with offshore siting and power transmission by marine cable back to shore, these costs are at least partly offset by the generally higher speed and more consistent winds offshore. Currently offshore wind projects generate power at around 4–6 p per kWh, but prices are falling as experience is gained and new technology emerges, with capital cost per kW installed falling by around 30 per cent over the past decade. Overall, despite the obvious problems of installing and maintaining devices in rough seas, offshore wind looks like becoming a major energy option, particularly as acceptable sites for on-land siting become harder to find.

Wave power

Wave power represents another major offshore energy resource. Waves are caused by the frictional effect of wind moving across the open sea – a series of rolling circular motions being set up under the surface. In effect waves are a form of stored wind energy. Typical energy densities at good sites (e.g. 100 miles or so off the north-west of Scotland) are on average around 50 kW per metre of wave front.

A series of floating wave energy converters, made up of large individual units linked together in chains, could be used to absorb some of this energy. Given a series of chains of total length 400 km, with a conversion efficiency of 30 per cent, the generating capacity would be 6 GW. A scheme of that scale would be theoretically possible off the North Atlantic coast of the UK and, if built, it would be equivalent to about 10 per cent of the UK's total current installed generation capacity. Indeed in principle there could be larger schemes.

The actual total wave energy potential in practice would obviously depend on how many chains of converters were installed, their overall conversion and transmission efficiencies and the location of the devices, but estimates for the UK range from 20 per cent of current UK electricity requirements up to 50 per cent or more, using sites off the north-east and east of Scotland, and to the west of Cornwall (Ross 1996).

Given its Atlantic positioning, the UK has most of the best sites and about a third of the total European wave energy resource. Even so there are significant resources elsewhere in Europe, for example off Ireland, Norway, Portugal and Spain. Similar resources exist elsewhere around the world, for example off Japan, Australia and North America.

Scale prototypes of some large conversion systems were tested in the UK in the 1970s, perhaps the simplest being a squeezed air-bag system (the Clam). See Figure 7.2. The most complex is Salter's gyro-stabilised 'nodding duck', a scale model version of which was tested in Loch Ness. Clearly, when operated out at sea, they would have to be engineered to withstand storms, but maintenance problems could be eased by towing individual units back to shore; the offshore position would attract few environmental siting constraints.

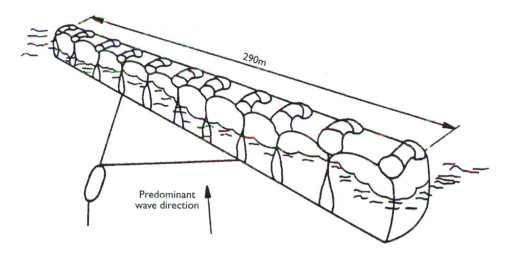

Figure 7.2 *The Sea Clam wave energy device. The wave action compresses a series of segmented air bags mounted on a fixed frame and the expelled air feeds to a turbine that generates power*
(*Source:* SEA-Lanchester, Coventry University/AEA Technology)

Smaller amounts of power, at possibly less cost, can be obtained from devices operated nearer to the shore ('in-shore' or 'coastal' systems) and from shore-mounted units, e.g. sited in gullies ('on-shore' systems). In the event, rather than large deep-sea projects, it has been smaller scale devices in these sorts of location that have gone ahead. Most of them make use of oscillating water columns, with, essentially, the rise and fall of the water, trapped in an inverted cup, squeezing air out an aperture at the top, and driving a two-way turbine. A small 75-kW on-shore system was constructed in a gully on Islay in Scotland, followed by a 500-kW device, called the Limpet, the first wave device to be connected to the grid. The same team had earlier developed a 2-MW in-shore device, the Osprey. However, this met with problems: during its initial test in 1995, the first prototype was damaged by storms before it could be secured properly on the sea-bed. The developers, Wavegen, are, however, pressing ahead with a second version.

Meanwhile, various other devices have been tested in Scottish waters, including Ocean Power Delivery's articulated 'sea snake', the Pelamis. This differs from earlier designs in that, rather than using oscillating water columns, it generates power from the snaking motion induced in a series of linked, floating cylinders, using hydraulic links between the segments. In addition, the device is tethered at the nose, so as to meet the wave head on, rather than flank side on, as with the Clam. Consequently, it is claimed to be likely to be more robust in storms.

Many other in-shore and on-shore devices have also been developed elsewhere, notably in Sweden, Norway, Denmark, the Netherlands, Portugal, Australia and Japan (see Box 7.3).

Box 7.3

Wave power around the world

Wave energy resources are largest where strong prevailing winds run over extensive seas, as in the North Atlantic. *Norway* was one of the first to get into wave energy, with the TAPCHAN, a tapered channel device built into a fjord, and the development of a pioneering, cliff-mounted, oscillating column device. Portugal also developed oscillating wave devices and installed at 500-kW device on the island of Pico in the Azores. The Dutch have developed a novel Archimedes Wave Swing device, consisting of a series of linked mushroom-shaped air chambers that rise and fall in sequence with the wave pattern, and pump air through a turbine to generate power.

Denmark's ambitious wave energy programme had an initial budget of DKK 20 m (£2 million) and aimed at repeating the very successful Danish wind energy programme by supporting a range of concepts and inventors. Several prototypes resulted, including the Wave Plane, the Point Absorber system, the Wave Dragon and the Swan. The Wave Plane is particularly interesting. It funnels incoming waves of varying frequency into a series of troughs to create a vortex which is used to drive a turbine. A 1:5 scale model has been under test in Mariager Fjord in Jutland since 1999, and scale versions of all the other devices have also been under test.

Although most of the development work so far has been European and geared to the North Atlantic, the Pacific also offers some good sites and in addition to developing small, oscillating water column devices mounted on breakwaters, *Japan* has developed a 110-kW Mighty Whale floating system. Developers in *Canada* and the *USA* have also shown interest in making use of wave technology. A new US company, AquaEnergy, is planning a 1-MW offshore demonstration project for near Neah Bay in Washington State, using *Swedish* wave pump technology. The project is expected to generate power at about 6 US cents per kWh, and if expanded to 100 MW, it is claimed that the cost could be as low as 4 to 5 US cents per kWh, similar to hydropower. Further north, BC Hydro is planning to develop up to 4 MW of ocean wave power as part of a broader plan to install 20 MW of renewable energy generation on Vancouver Island, British Columbia. It has chosen to use a shore-mounted wave energy device, featuring a parabolic wave focusing system, which has been developed by Energetech in *Australia*, following tests on a 500-kW-rated prototype at Port Kembla, New South Wales.

Although the total amount of power from individual in-shore and on-shore systems may be relatively small, the costs are lower than with the more complex deep-sea wave energy systems, and smaller scale wave power systems of this sort are seen as being likely to be of particular relevance to developing countries with suitable coastlines and to remote islands, where the only source of power might otherwise be imported fuel. However, although the individual systems may be relatively small, there are plans for creating wave parks with large numbers of wave devices feeding into a marine power cable. On this basis, wave energy could begin to make a major contribution. For example, Alla Weinstein, CEO of the US wave energy company AquaEnergy (see Box 7.3), has claimed that 'offshore wave power has the potential to satisfy 5 per cent to 10 per cent of total US power demand within 20 years'.

Tidal power

Tidal power and wave power are often confused. However, whereas wave energy is created by the winds, the tides are the result of the gravitational pull of the moon modified by that of the sun, the tidal height sometimes being increased by local funnelling effects in estuaries. The power that can be generated by a tidal barrage across a suitable estuary can be calculated in rough terms by assuming that the barrage traps a constant area A of water at high tide, which is then all passed through turbines, to the low tide level; falling through a distance r, the high to low tide range. The mass of water is dAr, where d is the density of water, and it would fall on average through $\frac{1}{2}r$. The potential energy per tide is thus $\frac{1}{2}dAr^2$.

This square law implies that estuaries with large ranges will pay dividends. Hence the interest in the Severn estuary in the UK where topographic funnelling effects produce a tidal range of up to 11 metres.

Since there are only two tides per day a tidal barrage will not operate continuously – typically they can supply power for say three to five hours per tide, depending on the operational pattern adopted (e.g. generating on the incoming flow, on the ebb or on both, using two-way turbines).

The proposed 11-mile long, 8.6-GW ebb generation Severn Tidal Barrage (see Figure 7.3) would, for example, generate 17 TWh per annum in total, around 6 per cent of annual UK electricity requirements, although the value of this would vary since not all of this would be available at times of peak demand.

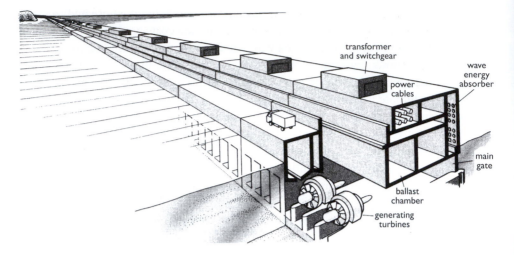

Figure 7.3 *Artist's impression of the Severn Tidal Barrage concept. The 11-mile long barrage would have 8.6 GW of turbine capacity and generate around 6 per cent of the UK's electricity, but it would cost around £10 billion to build*
(*Source:* Open University T362/T302)

This disadvantage can be offset to some extent by using barrages as a short-term pumped storage reservoir (e.g. using off-peak electricity from other plants to pump water behind the barrage ready for electricity generation via the barrage turbines when needed). In addition, if there are several barrages in different parts of a country (e.g. in the UK, on the Humber, Mersey, Dee, Solway Firth, etc.), the net output would be more nearly continuous, since the tides at each point occur at different times, the difference being up to five hours for some sites in the UK (Watson 1994).

The basic technology of power generation is similar to that used in **hydroelectric dams** – large turbines are mounted in vast concrete structures – although the head of water developed by tidal barrages is much less than that in conventional hydro projects.

However, with tidal barrages, there is no need to impound and flood vast new areas – this being one of the reasons why objections have increasingly emerged in relation to large hydro schemes. Nevertheless, proposals for tidal barrages have met with significant objections from environmentalists concerned about the likely negative impacts on the local ecosystem of altering the tidal range. Extensive studies have already been carried out on the Severn estuary, but a lot of careful environmental impact assessment work still needs to be done to resolve this issue.

One alternative to barrages across estuaries is to build complete free-standing lagoons in shallow water out to sea, filled by the incoming tide and emptied through turbines at low tide. In one version of this idea being planned by the US company Tidal Electric for testing at a coastal site off Wales, there would be two separate reservoirs within the lagoon, emptied in turn, so that more nearly continuous power could be produced.

Although the UK has an enviable tidal energy resource, put at around 20 per cent of electricity requirements if fully developed, there are also opportunities for tidal energy elsewhere in the world. A 250-MW tidal barrage already exists on the Rance estuary in Brittany and there are smaller barrages in the former Soviet Union, Canada and China. The total world tidal barrage potential is put at about 120 GW, which could produce around 190 TWh p.a. (Baker 1991).

Rather than having to build large expensive and environmentally invasive barrages, there is also the option of collecting energy by using submerged wind turbine-like devices from the fast-moving *tidal currents* that exist in channels offshore in some areas. Some prototypes have already been developed including a 10-kW tethered device, tested in a loch in Scotland in 1994 and a 300-kW 'seaflow' marine current turbine, mounted on a pile driven into the sea-bed off the north coast of Devon in 2002 (Figure 7.4). A novel sea-bed-mounted device, called the Stingray, was also tested in 2002, off the Shetlands' coast in Scotland. It consists of a hydroplane wing mounted on a hydraulic arm which rises and falls, extracting energy from the tidal flow.

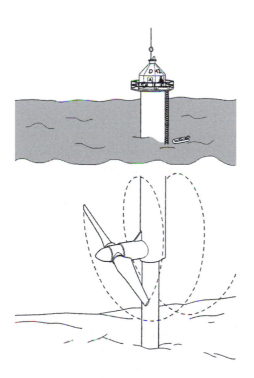

Figure 7.4 *'Seaflow' marine current turbine*
(*Source:* MCT)

According to a study by ETSU, in principle the UK might obtain up to 19 per cent of its electricity if all its tidal stream resources were tapped (Energy Technology Support Unit 1993a). Similar resources exist elsewhere, for example in the Straits of Messina in between Sicily and mainland Italy and tidal current systems could have very widespread applications around the world. Some could be very large. The Canadian company Blue Energy is developing an ambitious tidal fence, with an array of vertical axis water turbines mounted in a causeway between islands in the Philippines, which could ultimately be expanded to deliver 2,200 MW at full power. There have also been even more ambitious proposals, including the idea of using arrays of submerged turbines to tap the very large energy flows in the Gulf Stream, although these are not actually tidal, but are the result of the large-scale movement of water around the oceans driven ultimately by solar energy.

Energy crops

So far we have dealt with the use of natural energy flows, derived directly or indirectly from solar energy or lunar gravitation effects. However, solar energy also supports the growth of plant life and other forms of biomass, and this stored solar energy can also be seen as a renewable resource, as long as its rate of use does not exceed the replacement rate.

Growing fuel rather than food could well become a significant new option for farmers around the world. As long as the replanting rate matches the rate of use, the overall process of energy crop growing and combustion can be 'greenhouse neutral' in that plants absorb carbon dioxide while growing, so that no net carbon dioxide is produced.

There is a range of possible uses for 'bio-fuels' of various sorts – liquids (e.g. ethanol and bio-diesel) for transport use, and gases (e.g. methane) and solids (e.g. wood) for heating or electricity production. Ethanol has been produced from sugar cane in Brazil for some years. Bio-diesel (rape methyl ester) is now being produced in some parts of the European Union, especially in France, from oil seed rape. However, as we saw in Chapter 3, there are some questions as to the net energy ratio (energy input compared to energy output) given the energy requirements of harvesting and transport.

By contrast, with solid fuels from 'woody biomass', the energy ratio is believed to be higher. *Short rotation arable coppicing*, e.g. using fast-growing willows or poplars, is currently seen as likely to be an important source of fuel for electricity generation in the UK – indeed the UK Government's Department of Trade and Industry (DTI) has estimated, perhaps rather optimistically, that the maximum total UK resource potential by 2025 could be up to 150 TWh p.a. – nearly one-half of

current UK electricity requirements (Department of Trade and Industry
1994b).

Short rotation coppicing (SRC) does not involve full tree development
– instead the thin willowy growths, reaching perhaps 15 feet, are cut
back to stumps every three to five years. Large areas could be involved
with mechanised cutters passing periodically along corridors through
the coppice plantations. The resultant wood chips would be transported
periodically to combustion plants. The combustion process can be very
efficient, especially given the development of advanced co-generation/
CHP techniques. For example, for gasified biomass used to power
steam-injected turbines (the so-called BIG-STIG technology), operated
in the combined heat and power mode, conversion efficiencies of up to
80 per cent can be obtained.

The UK's first main venture in this field was the 10-MW ARBRE wood
chip-fuelled plant in Yorkshire, with wood gasification technology linked
to an 8-MW combined cycle gas turbine, with 2 MW of heat also being
available. The ARBRE project was very much a prototype, and in 2002 it
ran into problems with development funding after teething problems
delayed full commissioning. Clearly, the technology still needs to be
developed in order to reduce costs, and improve reliability. In addition,
there remain some environmental concerns with the energy crop side, for
example, in relation to the run-off of pesticides and herbicides, the impact
on wildlife and local disruption from the extra traffic movements required
for moving the resultant wood chips. There is also the problem of
emissions from the combustion of the chips.

The proponents of SRC argue that herbicides would only be used in the
first year of the cycle and fewer pesticides would be needed than for
conventional arable crops. They also claim that coppices tend to mop up
pollutants, thus reducing run-off into water courses. Coppices would be
located mainly on marginal land and would offer habitats, for example,
for birds and insects, especially given the open access corridors that
would traverse the plantations. Biodiversity could be ensured by using
multi-cloned plants or by mixing species: coppices have already been
shown to attract woodland animal species. Overall, if degraded or
abandoned cropland were used, the environmental and wildlife impact
would be positive.

The proponents of SRC argue that there would not be excessive transport
requirements. Wood chips would be stored on site at farms and
transported annually to combustion units on local industrial sites for

combustion. For, say, a 10-hectare coppice plantation, with 5 hectares harvested annually, only 75 tonnes of wood chip would need to be transported annually. This would involve 4 × 20 tonne loads or 8 × 10 tonne loads – not a huge annual volume. And it would replace the transport of the crops that might have been grown otherwise.

On balance, arable coppicing seems to offer significant advantages. However, there are clearly some unknowns: what exactly are the environmental and pollution implications; how economic will it be; how will it be planned; what scale is appropriate? Is it just an extension of conventional farming or something new?

At the very least, some form of planning control would seem necessary to avoid land use conflicts and visual disruption. Even so, energy crops could well become a major new energy source.

Biomass wastes

The idea of producing specially grown energy crops is relatively new, but the various types of biomass waste materials already produced in modern societies also represent a useful source of energy. For example, there is a whole range of farm wastes including pig slurry and chicken manure. Some animal and agricultural wastes can be converted into methane gas, via anaerobic digestion techniques, while others can simply be burnt as a fuel: there are now power plants in the UK fuelled with chicken droppings. Residual wood from forestry operations can also be used as a fuel, as can straw.

In addition to these agricultural sources, there is a further source of biomass available in industrial societies in the form of domestic and industrial waste. *Energy from Waste* projects of various types are already widely established around the world. Electricity generation using waste combustion is a well-developed option, although, as projects have spread there have been concerns raised by environmentalists and local residents about toxic emissions, such as dioxins, from the combustion of plastics. Dioxins are cancer forming at very low levels of exposure, but the plant operators claim that the current generation of combustion technology can keep emissions well below danger levels. Their opponents are, however, not convinced and are also worried about acid emissions. See Box 7.4.

Improved energy conversion technology, for example, the adoption-fluidised bed combustion may help to reduce acid emissions, while

Box 7.4

Dioxins and acid emissions from waste combustion

Dioxins are polychlorinated compounds, which can be formed, along with variants including furans, when compounds containing chlorine are burned in domestic refuse and industrial waste incinerators. They can be found in both the flue gases and the fly ash. They can also be produced from other combustion processes, including bonfires. There are 210 different types of dioxin and furans, of which 17 are toxicologically significant. There has been an extensive debate on the scale of health risks associated with the combustion of municipal solid waste (MSW).

While according the Porteous (2000), many old, poorly designed MSWaste incinerators discharged dioxin at levels greater than 45 nanogrammes TEQ (toxicological equivalent) per normal cubic metre (45 ng TEQ/Nm^3), the dioxin release levels from modern, state-of-the-art, energy-from-waste (EfW) plants are now in the range 0.02–0.4 ng TEQ/Nm^3. He notes that 'Under UK Environment Agency regulations older plants which could not meet the new standard of <1.0 ng TEQ/Nm^3 were closed down. EC requirements are <0.1 ng TEQ/Nm^3 which all new plants can readily meet. These standards imply a 98 per cent reduction in dioxin emissions from MSW combustion'.

However, environmentalists in organisations like Greenpeace and Friends of the Earth are not convinced by these arguments. They claim that in practice emission controls may not always be operated correctly and that in any case the current safety levels are too high. Certainly, a study by R. De Fre and M. Wevers ('Underestimation in dioxin inventories', *Organohalogen Compounds*, vol. 36, 1998) suggested that the regulatory monitoring regime is underestimating dioxin emissions by a factor of 30 to 50.

On the other side of the debate, it has been argued, as Warmer Bulletin 51 (November 1996) put it, that 'modern Energy from Waste units can actually operate as dioxin sinks, trapping and destroying more of the dioxins which were present in the fuel input than they generate as by-products of combustion'.

This is an area where scientific analysis and pressure groups campaigning often both collide and collude – there seems to be rival data on each side. For example, one US study used by campaigners suggested that 7 per cent of cancers were due to dioxin exposure, and that around 80 per cent of the dioxin exposure is from incinerators. However, the official view in the UK is that incineration of MSW makes a much smaller contribution than other sources; for example, a 1998 Environment Agency study put the figure for incinerators at 1 per cent of total dioxin intake by human beings, whereas car exhausts contribute 9.5 per cent. Other large emitters include the steel industry and domestic bonfires.

The hardline position is that dioxin production should be reduced to zero; and on this basis Greenpeace has been campaigning against the production of plastics like PVC

continued

(polyvinyl chloride), a material that is, of course, ubiquitous in many consumer products. The use of bottom ash residues from MSW combustion as building materials or for paths has also been seen as a problem.

However, it is not just dioxins that environmentalists point to, there is also the impact of *acid emissions*. For example, in their campaign in 2000–1 against proposals to build up to seventy new waste combustion plants around the UK, Greenpeace made use of a DETR prediction that 50 tonnes of nitrous oxide emissions would on average lead to one death, to calculate that, despite using newer, cleaner technology than the current incinerators, the new incinerators would release 17,000 tonnes, which might be expected to 'cause at least 350 deaths a year' for the 25-year lifetime of the plants – 8,700 deaths in all (*Observer* 21 October 2000).

Greenpeace noted that

> the Department of Health's Economic Appraisal of Health Effects of Air Pollution (EAHEAP) have prepared a cost benefit analysis which calculated 'Number of deaths not brought forward' per tonne of pollution avoided. The figures are 0.02 deaths a year per tonne of NOx, 0.005 deaths per tonne of SO_2 and 0.002 deaths per tonne of particulates. The figures are quoted in the DETR's report: Regulatory and Environmental Impact Assessment of the Proposed Waste Incineration Directive.

They noted that Environment Minister Michael Meacher said in a House of Lords Inquiry in April 1999 'the emissions from incinerator processes are extremely toxic. Some of the emissions are carcinogenic. We know, scientifically, that there is no safe threshold below which we can allow such emissions. We must use every reasonable instrument to eliminate them altogether'.

Clearly, the debate remains a live one, and it has certainly given rise to some colourful rhetoric. For example, Rob Gueterbock, a Greenpeace campaigner who at one stage scaled an incinerator chimney in protest, said: 'With incineration your rubbish comes back to haunt you. You put it out through your door and it comes back as air pollution through your window.' On the other side, it has been noted by Prof. Porteous that the grammes per volume emission requirements for a new Swedish plant, 0.1 ng/m^3, was equivalent to 'a quarter of a standard 3g sugar lump dissolved in Loch Ness' (Porteous 2000).

pyrolysis and gasification techniques are claimed to reduce dioxin production, and yield liquids or gases, which can be used as a fuel for electricity production, or as heating fuels. However, these are as yet underdeveloped technologies and it is not clear either whether they will be economic or meet the concerns expressed by environmentalists.

The collection of methane gas from landfill sites is another option for obtaining energy from wastes. There can be problems with toxic materials leaching out from landfill sites into the local water courses, but, since landfill sites exist in large numbers it would seem better to gather the gas they produce, rather than letting it leak out and contribute to global warming. This also reduces the risk of explosions at the landfill site.

However, not everyone is convinced that waste should be seen as a genuinely 'natural' or renewable source of energy. Certainly, large amounts of industrial and domestic wastes are produced regularly in industrial societies, but most environmentalists argue that the generation of material waste (for example, from packaging) should be minimised in the first place, and that the energy content in whatever is left can be recovered more effectively by materials recycling schemes (Greenpeace 1992).

On this view, direct energy recovery, via combustion or landfill gas, should only be the last resort. The counter viewpoint is that recycling can be energy intensive and that, even with a major commitment to waste minimisation and recycling, there will still be a significant amount of waste, representing a large potential source of energy (Porteous 1992). However, there are also wider environmental issues to be considered, for example, the ecological value of the waste material. Combustion inevitably destroys valuable organic material that previously was returned back to the ecosystem via various routes. If however, the refuse from cities is collected and, in effect, sterilised by combustion, we will have to replace the valuable organic material it contained with artificially produced fertilizers, the production of which may use more energy than is recovered from burning the waste (Clarke and Elliott 2002).

The 'dream solution' for wastes is to avoid both combustion and landfilling and the resultant risk of dangerous emissions. Biogas production from farm and animal wastes by anaerobic digestion is well established, particularly in warmer countries, and generates fertilizer as well as methane gas. In addition, advanced energy conversion techniques (some using biological processes) are emerging, which could lead to commercial generation of hydrogen gas from wastes. This could be used for heating or for electricity production, for example, via a fuel cell – a device that runs something like electrolysis in reverse, converting hydrogen to electricity. That could even offer the possibility of urban vehicles being run on hydrogen produced from urban refuse.

Other renewables

There are many other sources of renewable energy ranging from giant schemes for making use of the temperature differentials between the surface and the depths of the oceans (the so-called 'Ocean Thermal Energy Conversion' technology or OTEC) to the use of geothermal heat deep underground. OTEC has yet to be developed seriously, and looks likely to be expensive, but geothermal energy is being used widely around the world.

Extracting heat from *geothermal* sources deep underground is not strictly a continuously renewable process: the heat gradient is gradually diminished, although it may be replenished later, if heat extraction is halted for some years. Natural hot water and steam geysers exist in various parts of the world, for example in the USA, and the UK has a number of hot 'spa water' sites. This type of geothermal heat comes from aquifers near to the surface, and this resource is tapped for hot water or power production in many countries, such as Japan and the USA. It is a quite significant resource: in addition to providing around 34 TWh of direct heat supplies, there is a total of nearly 7 GW of geothermally powered electricity-generating capacity in place around the world.

An even larger energy resource can be tapped if access can be gained to the so-called 'hot rocks' deep underground. The technology for deep well

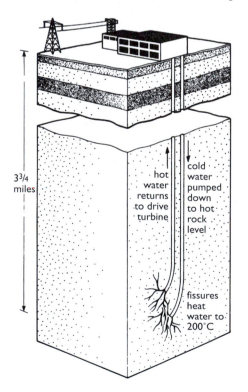

heat extraction from 'hot dry rocks' typically involves drilling two bore holes to a depth of around three to four miles in suitable locations, and then forming a series of geological fractures in the rocks, to connect up the bottom ends of the bore holes. Water is then injected under pressure down one bore hole so that it percolates through the fracture pattern, which acts as a heat exchanger, to re-emerge up the other bore hole as steam. This can then be used to drive turbines for power generation (see Figure 7.5).

Figure 7.5 *Artist's impression of a geothermal 'hot dry rock' well*
(*Source:* Open University T362)

Heat pump technology is also sometimes categorised as relying on geothermal heat sources. Certainly, heat collected via pipes buried in the soil can be pumped to warm buildings – the heat pump system in effect working like a refrigerator in reverse, pumping heat from outside into radiators. However, the heat in the ground is mainly from ambient sources, that is, solar heat. At the depths used for most ground-sourced heat pumps, there can hardly be much of a geothermal contribution. Heat pumps also need energy to run. However, driving a heat pump can be a very efficient use of energy, multiplying the value of the energy used by perhaps up to three or four times.

Finally, there is *hydroelectric* power, which has been exploited on both a large and small scale for many years. Large hydros are major suppliers of energy – there is around 650 GW of installed capacity in place, mostly in 300 large projects. However, in recent years, there have been social and environmental concerns about large hydros, and some new projects have met with opposition. One key issue has been the problem of social dislocation caused by having to move people from the areas being inundated to create reservoirs. In addition, it seems that in some locations (e.g. in warm climates) the anaerobic digestion of biomass brought continually downstream and tapped by the dam can create methane gas to such an extent that a coal-fired plant of the same capacity would produce less net greenhouse gas impact (World Commission on Dams 2001). What we are seeing here is the result of the disruption of a natural energy flow, which previously ensured the continual agitation of the water, so that anaerobic processes were minimised. There may be solutions, including the redesign of dams so that some water flows are redirected to stir up continually the 'sump' in front of the dam, and the scale of this problem will in any case depend on the local topography and ecology.

While some large hydros may have problems, smaller scale hydros – as in so-called 'mini- or micro-hydro' projects in rivers and streams – continue to be developed successfully around the world. There are 50,000 small hydro power stations in China alone. The individual generation capacities involved are small, but taken together they represent a significant amount of power, and this option is particularly relevant in developing countries, providing cheap local power inputs. Micro-hydros also avoid most of the environmental problems associated with massive hydro schemes. Certainly, small projects are less invasive locally.

The total renewable resource

Having now reviewed some of the key technical options, what is the overall scale of the renewable energy resource? The simple message is that the basic energy source is very large. The sun delivers a vast amount of energy to the earth – much more than could ever be used, and the resultant natural energy flows represent a very large, non-finite resource base, with the solar radiation input alone amounting to around 90,000 TW. For comparison, total global energy consumption in 1990, measured on a continuous equivalent power basis, was only 13.5 TW.

Not all of this 90,000 TW can be successfully captured and used. Most of the flows are diffuse, some are intermittent, and the efficiency of conversion technologies has to be taken into account, as does the location of the source.

Projections as to the amount of useful energy that can be extracted in practice vary, but Jackson, reviewing a series of papers covering the field in *Energy Policy* journal, put the world's recoverable tidal resource at 0.1 TW, the hydro resource at 1.5–2 TW, the wave resource at 0.5–1 TW, the wind energy resource at 10 TW, and the solar resource at 1,000 TW (Jackson 1992).

Of course, what matters is the extent to which it can be developed in practical terms, and the timescale on which this might be achieved. That of course will depend on, amongst other things, the success of the technological development programme, the economics and the environmental constraints, and, crucially, for a new area of technology, on the level of support given. We will be looking at the state of play around the world in the next chapter, after which we will be discussing some of the strategic development issues.

Summary points

- Most renewable energy sources derive directly or indirectly from the sun, the exceptions being tidal power and geothermal energy.

- Of the 'new' renewables, wind power has been developed quite widely, as has solar power. Wave and tidal energy are less widely used so far but energy crops are beginning to become established as a major new option. PV solar could be next in line.

- The renewable energy resource is very large, but only a small part of it can be tapped effectively and economically.

- Nevertheless, if fully developed, the use of renewable sources could, in principle, meet all the world's energy requirements.

- Renewables are still only a relatively marginal source, but their use is growing around the world.

Further reading

The Open University course text G. Boyle (ed.) *Renewable Energy: Power for a Sustainable Future*, co-published with Oxford University Press in 1996 and updated in 2003, provides an excellent introduction to the technology, with the emphasis on the UK.

The classic (but massive) text is Bent Sorenson's *Renewable Energy* (Academic Press, London, 1979, second edn 2001). John Twidell and David Wier's *Renewable Energy Resources* (E. & F.N. Spon, London, 1990), is a shorter text, providing a good introduction to the basic physics involved. A useful US source is Chris Flavin's *Power Surge: Guide to the Coming Energy Revolution* (Worldwatch Institute, Washington DC, 1995).

There are a number of useful books on specific technologies including Paul Gipe's very readable *Windpower Comes of Age* (Wiley, Chichester, 1995), the more academic *Wind Energy in the 21st Century: Economics, Policy, Technology, and the Changing Electricity Industry* by Robert Y. Redlinger, Per Dannemand Andersen and Poul Eric Morthorst (Palgrave/UNEP, Basingstoke, 2002), Clive Baker's textbook *Tidal Power* (Peter Peregrinus/IEE, London, 1991) and the seminal political history of *Power from the Waves* (Oxford University Press, Oxford, 1995) by David Ross.

In such a rapidly moving field, web-sites are invaluable. One of the best on wind power is: http://www.windpower.org; for solar power see: http://www.ises.org; for wave power see: http:// www.wave-energy.net.

8 Renewables worldwide

- UK renewables
- US renewables
- European renewables
- Japanese renewables

Renewable energy technology has developed rapidly in recent years and it is being used widely around the world. This chapter explores the state of play so far, focusing on developments in the UK, Europe, USA and Japan, and looks at the reasons why some countries are pushing ahead with renewables urgently.

Renewables old and new

The use of renewable energy is not new. As we noted earlier, biomass in the form of firewood still represents the main fuel source for many of the world's people, and conventional large-scale hydroelectric generation represents a major existing use of a renewable energy source. However, wood fuel is becoming increasingly scarce, and given the high capital cost of large hydro projects and, in some cases, major environmental impacts of such schemes, there has been growing interest in smaller scale hydro plants around the world.

Of the 'new' renewables, the use of solar energy for space and water heating obviously has attractions in many countries. There are major projects underway around the world for electricity production, either using solar heat to raise steam, or using sunlight to power photovoltaic cells. However, in terms of electricity production, it has been wind power that has made the biggest impact: as noted earlier, by 2002 there was more than 31 GW of wind turbine capacity in plants installed around the world.

While the review of international developments that follows focuses on Europe, the USA and Japan, it should be noted that renewable energy technology is also being deployed quite widely elsewhere. India and China both have major wind power programmes, and in many developing countries the use of solar energy is an obvious option. There are around 2 billion people in the world who have no hope of getting access to grid electricity and for whom PV solar energy may be an excellent option. In addition, some Middle Eastern countries are interested in exploring the idea of hydrogen production powered by photovoltaics as a long-term alternative to reliance on oil sales. For example, Saudi Arabia is collaborating with Germany on a 'Hysolar' PV project. For the moment, though, as this example illustrates, most of the running is being made by the 'developed' countries. However, this situation may not continue. For example, China already manufactures more PV cells than the UK, and some of the newly emerging Asian economies could well take over a leading role in the renewable field in the years ahead.

With access to fossil fuels, particularly oil and gas, being constrained, the hunt is clearly on for alternatives. Much of the technological and economic history of the industrial world has been shaped by the geographical distribution of fossil fuel reserves. Indeed, modern geopolitics continues to be influenced strongly by the location of oil and, more recently, gas. In the same way, in the future, those countries with large-scale indigenous renewable resources and the technical capacity, financial means and political will to develop them may become dominant.

As Rodney Chase, Deputy Group Chief Executive Officer and Chief Executive of BP Oil, commented: 'It could well be that the first country to seriously address the issues of creating a market for renewables would become the central location for a major new international business sector – with all the positive consequences that carries in terms of economic activity and employment' (Chase 1998).

Developing renewables in the UK

As a hilly country surrounded by stormy seas, the UK has quite large renewable energy resources, possibly the largest per capita in the world. For example, the overall UK resource potential for electricity-producing renewables, generating at a cost of less than 10p per kWh (and assuming an 8 per cent discount rate on investment), has been estimated by the UK Government as being up to 1100 TWh p.a. by the year 2025 – more than three times the annual national demand for electricity (DTI 1994b).

This of course is only the theoretical maximum, ignoring many practical, economic and siting constraints. A more realistic, fairly conservative, mid-range estimate, adopted by the Government's Renewable Energy Advisory Group, was a 20 per cent contribution to UK electricity requirements by 2025, with around 10,000 MW (net) of renewable capacity generating about 60 TWh p.a. (REAG 1992).

Even so, the UK clearly has a very significant renewable energy resource. This first became apparent when ETSU, the Government's Energy Technology Support Unit, started to assess the renewable options.

ETSU was set up following the oil crisis in 1974, and it oversaw and assessed a wide range of government-funded research and development projects in the renewable energy field. Wave power, tidal power and geothermal energy were the initially favoured options but, from around 1984 onwards, wind power and, more recently, energy crops came to the fore. We will be looking at some aspects of this programme in later chapters.

By 1994, a cumulative total of around £232 million had been allocated. However, after reaching a peak of around £25 million in 1994–5, the level of R & D support began to fall. The Conservative Government argued that the R & D programme had successfully identified those options that were likely to be commercially viable – namely on-land wind, waste combustion and some biofuels (Department of Trade and Industry 1994b). This strategy of focusing on the currently commercial options had its limits: the renewable energy field is very new and it could be seen as rather short-sighted to cut funding for the longer term projects (Elliott 1996). The Labour administration elected in 1997 reversed some of these trends and has allocated increasing amounts of money to the newer renewables, particularly offshore wind, energy crops and, to a lesser extent, PV solar, wave and tidal energy.

The Non-Fossil Fuel Obligation

Although the level of R & D has been rather low in the UK, governments of varying political hue have accepted that some interim support was needed for those renewable energy technologies that were near to, but not quite yet at, the commercial lift-off point. In order to support the commercialisation process, in 1990 the Conservative Government introduced a special interim *Non-Fossil Fuel Obligation (NFFO)* – a cross subsidy arrangement funded by a small levy (initially less than 1 per cent) on fossil fuel supplies, with the extra cost being passed on to electricity consumers. This levy was actually part of a larger scheme introduced primarily in order to support nuclear power. The so-called 'fossil fuel levy', of around 10 per cent, raised around £1.3 billion each year from consumers, most of which went to offset the cost of nuclear generation. The renewables only obtained a very small part: £30 million p.a. in the first phase of the scheme, although this gradually increased, and reached a peak of £127 million in 1998–9.

The nuclear part of the NFFO was set at the entire output of the nuclear plants in England and Wales (around 8.5 GW): separate but similar arrangements were made in Scotland. Following the intervention of the European Commission, who were concerned about an unfair subsidy being provided to nuclear in the UK, an agreement was reached that the nuclear part of the scheme would be wound up by 1998.

The first renewable NFFO round was in 1990 and was set at 102 MW declared net capacity (as you may recall from Chapter 7, Box 7.2, the 'declared net capacity' is the capacity rating adjusted to take account of the fact that some renewable energy technologies make use of intermittent natural energy sources like the wind).

Subsequently, four further rounds of the renewable NFFO have emerged up to 1999, and similar schemes were put in place in Scotland and Northern Ireland. Developers were required to submit bids for each round of the NFFO and the sister schemes, and there was a price versus capacity conflation process, overseen by a government agency, to select those with the lowest price bids up to the total capacity set for each round. So there was a strong competitive element in the process. The regional electricity companies had to take all the power specified in each round but could pass on the extra cost of using renewables to their customers. This led to price increases of less than 1 per cent, lost in the larger nuclear part of the levy, until its demise in 1998. The overall target for the renewable part of the NFFO was set at 1,500 MW declared net capacity by the year 2000 (DTI 1994b), although in the event less than 1,000 MW was actually achieved by that time, mainly due to problems with obtaining local planning consent for wind projects and waste combustion projects.

The Renewables Obligation

In 2002 a new scheme replaced the NFFO – the *Renewables Obligation*. A new system was necessary following the restructuring of the electricity market. Under the Obligation, electricity supply companies had to move in stages so that by 2010 they would be obtaining 10 per cent of their power from renewable sources. This was in line with the national target set by the Labour Government of obtaining 10 per cent of UK electricity from renewables by 2010.

In the run-up to these targets, interim Renewable Obligation targets were set, and, when they complied with what was required, companies would be given Renewable Obligation Certificates (ROCs) on a MWh basis. As with the NFFO, the companies could pass the extra cost of compliance on to consumers, and this was expected to involve a price rise of around 4–5 per cent by 2010. However, a price ceiling was in effect put in place by the provision of escape routes for companies unable to meet the Obligation. They could either pay a 'buy-out' fine of 3p per kWh, or buy in Renewable Obligation Certificates from companies who had been able to earn more than they needed. The result was that, buying in renewable power would be economically attractive if it cost less than 3p per kWh above the price of conventional power (around 2p per kWh when the Obligation was first conceived). The 'buy-out' price of 3p per kWh thus set the value of the tradable Renewable Obligation Certificates, and also set a ceiling, of around 5p per kWh, for renewable generation, depending on the current level of conventional electricity prices. Similar arrangements were made in Scotland.

In parallel, the government introduced a Climate Change Levy, a surcharge on the business use of energy, set at 0.43p per kWh for electricity. Some of the larger energy users were offered concessions, in return for agreements on carbon emission-reducing strategies, and renewably sourced power was exempted from the levy. The result was that most of the renewable electricity supplies (amounting to around 6.8 TWh in the first year of the scheme, 2001–2) were snapped up by companies keen to avoid the levy. Indeed the scale of the demand seems to have outstripped the supply, with energy supply companies having to turn them away. The problem was that there was no obligation of generators to invest in new capacity to supply the energy retail market. That was left to conventional supply and demand economics, and, despite the demand created by Renewables Obligation and Climate Change Levy, it has taken time for investors to decide to move into what some evidently saw as a risky market.

Gradually, however, investment has begun to flow, and it is expected to reach £1–2 billion by 2010. The main signs of interest have been in offshore wind projects, but other smaller projects are also being taken forward. Even so, it does seem rather odd that the UK, with arguably the world's best renewable resource, has chosen support mechanisms, the NFFO and then the Renwables Obligation, which has meant that it was exploited only very slowly. While by 2002 the UK has only managed to install around 500 MW of on-land wind capacity, Germany, where a different support system was used, has installed 10,000 MW, despite having a much worse wind regime. Meanwhile, Denmark, which had started its renewable energy programme before Germany, had reached the point where up to 18 per cent of its electricity was being supplied from wind projects. It is worth exploring the reasons for these large differences in a little more detail.

Price versus capacity: the European experience

As we have seen, in the UK, the NFFO required electricity companies to buy in specified amounts of renewable energy, with a competition being held amongst rival candidate projects to meet this demand. A series of separate 'technology bands' were established, for wind, biofuels and so on, to segment this competition – so wind would compete with wind and so forth.

The result of this competitive approach was that prices began to fall dramatically. Whereas wind projects got 11p per kWh in the 1991 NFFO competition, by 1998 wind projects were going ahead at 2p per kWh in some locations – cheaper at that time than gas CCGTs. However, as we have also seen, the amount of renewable energy generating capacity that was installed was small compared, for example, with Germany.

In Germany, and also in Denmark until 2001, quite generous support was provided under the so-called Renewable Energy Feed-in Tariff scheme (REFIT), a quota system that guaranteed prices to generators – up to 8 US cents per kWh, depending on the source. The result was a boom in construction of wind projects, which has continued under variants of the REFIT system – the Electricity Feed-in Law (EFL) having been replaced by the 'Erneuerbare Energien Gesetz' (EEG) (Connor 2001).

However, the difference between the UK and Germany is not just that the latter had, in effect, higher subsidies. The highly competitive system used in the NFFO, and in particular the tight contracts in the early years due to the 1998 deadline for NFFO1 and 2, meant that wind developers had to select very high wind speed sites (7 m/s typically), most of which were in environmentally sensitive upland areas. In some cases this provoked a major local backlash against projects. Local objections have meant that around 70 per cent of wind farm proposals have been turned down by planning bodies. The improved economics of wind turbines should now reduce the need to use sensitive and provocative sites, but a negative mood has been established in some areas which is proving hard to change. We will be returning to this issue in later chapters.

This brief history of UK developments raises a key strategic issue. Should we go for generation capacity, at whatever the initial cost, as Germany and some other EU countries have done, on the assumption that costs will come down later? Or should we get the prices down before expansion occurs, as the UK has done – even if expansion does not yet seem to be occurring?

Clearly, renewables like wind needed subsidies to get established, but at some point they should be able to compete unaided. Wind has nearly reached that point in some parts of the UK, and some waste/biofuel combustion options have already passed it. So you could say that, for these technologies, the NFFO has achieved its goal, even if it has not led to much capacity being installed.

However, there are new renewable energy options which need support. As we have seen, the UK has now abandoned the NFFO and replaced it by a new support mechanism, the Renewables Obligation (RO), with, in effect a 5p per kWh ceiling (at then current prices), for renewable power. A ceiling of 5p per kWh may be high enough to allow some offshore wind projects to go ahead, but to try to help support new projects the government decided to provide some extra funds (£200 million in all) in the form of a competitive series of grants to help new renewables like offshore wind, and also energy crop projects – for example, short rotation arable coppice, feeding wood chips to combined cycle turbines. The latter were offered around 8p per kWh (under old NFFO contracts), but it was hoped that ultimately they would get down to 3p per kWh.

The Renewables Obligation, like the NFFO, has a strong competitive mechanism built in, along with a Renewable Obligation Certificate trading option. It should keep generation prices down, but may not lead to enough capacity being installed to meet the 10 per cent by 2010 target. Clearly, the RO has created demand for new renewable capacity, but, even if investment is forthcoming for new generation projects, as seems to be happening, there may not be sufficient at below the 5p per kWh price level. Moreover, there would be even less available if this ceiling price fell, as conventional electricity prices fell, as has in fact happened. In which case, if they cannot find ROCs to trade, it is not clear what will happen – will the supply companies simply pass the fines on to the consumers? Or will the 3p per kWh 'buy-out' price be raised?

The RO does very little to stimulate investment in the newer renewables like wave and tidal current energy. Prototype systems are now being tested, with, at long last, some government support – but only £6 million in grants. The device teams hope to generate at around 7p per kWh and expect that to fall to 3–4p per kWh later, when mass-produced commercial-scale devices are developed. At that point they will be ready for take-up within the RO. But it will take time and money to reach this point, so they may not be able to contribute much to the UK's target of obtaining 10 per cent of its electricity from renewables by 2010.

As Table 8.1 indicates, many other European countries have higher targets, in part, to be fair, because the UK started from such a low level of renewable energy capacity, essentially a few hundred megawatts of hydro. By contrast Austria, Sweden and Finland enjoy very large hydro and traditional biomass contributions. As a result, in 1995, Austria obtained around 24 per cent of its total primary energy from renewable resources, Sweden 25 per cent, Finland 21 per cent, while the figure for the UK was just 0.7 per cent.

Even leaving large hydro aside, the UK is still some way behind the majority. As can be seen from the EU targets for

Table 8.1 *EU Directive: 2010 targets*

	Electricity from renewables (%) (excluding large hydro)
Denmark	29.0
Finland	21.7
Portugal	21.5
Austria	21.1
Spain	17.5
Sweden	15.7
Italy	14.9
Greece	14.5
The Netherlands	12.0
Ireland	11.7
Germany	10.3
UK	9.3
France	8.9
Belgium	5.8
Luxembourg	5.7
EU 15	12.5

Note

For comparison, the 2010 target for total EU renewable electricity *including* large hydro was 22.0 per cent. On this basis the UK's target was a 10.0 per cent contribution by 2010, with France overtaking the UK at 21.0 per cent, while Austria becomes the leader, at 78.1 per cent.

electricity from renewables, as agreed in 2002, despite its very large renewable resource, and the fact that this data leaves out large hydro, the UK still occupies a lowly position in this ranking. That seems to be the price paid for the emphasis on competition.

The funding arrangements in some other EU countries, such as the REFIT system, may have their problems, in terms of loading up utilities and consumers with extra costs, but competitive pressures are not absent. Generators who can make use of cost-effective equipment will still be better placed than those that do not. Despite the REFIT-type schemes in Germany and, until recently, Denmark, there seems to have been no lack of technological innovation – quite the opposite. Both countries have been at the forefront of wind technology innovation. The same can hardly be said of the UK.

Competition is not necessarily a bad thing. It can ensure that resources are used efficiently, and we need to increase energy productivity for environmental as well as economic reasons. But obsession with obtaining low prices can be counterproductive when it comes to developing new energy technologies.

The UK wave energy programme in the 1970s was halted because it was claimed that the price of electricity from this new technology would be very high – figures of 20p per kWh or more were bandied about. Solar photovoltaics has also been challenged on the same sort of grounds over the years. However, it makes little sense to try to decide on longer term energy options using early estimates of costs based on prototypes, or first generation technologies. Prices fall as the technology improves – assuming someone has the courage or foresight to fund its continued development.

As was noted in Chapter 3, there are well-established 'learning curves' showing how technologies improve in performance and cost over time (McDonald and Schrattenholzer 2001). The rate of improvement can perhaps be improved with more money and effort and support for consolidating a market for the power. But one thing is clear: without sufficient support you do not get any improvement, or much capacity. See Box 8.1.

Most of the rest of Europe seems to have learnt this lesson. Certainly, overall, the pace of renewable energy development in Europe is very rapid, especially given the EU's strong commitment of fighting climate change via the Kyoto treaty. Box 8.2 gives a brief snapshot of what is happening in the various countries, including those outside the EU.

Box 8.1

REFIT works best

The guaranteed price *Renewable Energy Feed-in Tariff* system (REFIT) used in Germany, versions of which have also be adopted in some other EU countries, including, until recently, Denmark, has provided significant subsidies for renewables. In Germany, payments to renewable energy providers in 2001 under the REFIT system, or rather its new variant the EEG, totalled 1.5 billion euros ($1.49 billion), 0.4 billion more than paid out in 2000, for 18 billion kilowatt hours (kWh) of power at 8.6 euro cents per kWh.

A report by the *European Environment Agency*, published in 2002, compared the successes and failures of EU renewable energy programmes between 1993 and 1999. It noted that three countries that guaranteed purchase prices of wind-generated electricity – Germany, Denmark and Spain – contributed 80 per cent of new EU wind energy output during the period. This suggests that feed-in laws work better than the competitive tendering mechanism adopted by Ireland and the UK, a point reinforced by the problems being experienced in Denmark when it switched from a feed-in tariff to a certificate trading system (which was later abandoned).

A report by the World Wind Energy Association noted that more than 80 per cent (1,144 MW) of the 1,388 MW installed around the world in the first half of 2002 were installed in three countries with guaranteed minimum prices: Germany, Italy and Spain. In countries with quota/certificate systems, including the UK, USA, the Netherlands, and, more recently Denmark, only 75 MW were installed. France and Brazil have decided to introduce minimum price systems that recognise the success of this framework. In the USA, the national environmental group, the Sierra Club has also been campaigning for Electricity Feed Laws for the USA.

The REFIT system has not been without its opponents – the European Commission would clearly prefer a market-led, Europe-wide certificate trading system, and that may well come about in time. Certainly, REFIT-type subsidies should not need to be retained across the board forever. They are useful at the early stage of a technology's commercial history, but they can be progressively withdrawn as it matures. Some have argued that tradable certificates can be just as effective in providing support, but so far, as we have seen, this has not been the reality. If a fully developed certificate trading system evolves – possibly a pan-European one, following the planned (or rather sought for) harmonisation of the EU energy market – then perhaps matters might be different. But for the moment, a stepped/phased REFIT system seems like the better option.

Sources: 'Renewable energies: success stories', Environmental Issue Report 27, Ecotec Research and Consulting Ltd and A. Mourelatou, for the European Environment Agency, Copenhagen, 2001
http://reports.eea.eu.int/environmental_issue_report_2001_27/en

EIGreen Project 'Action plan for a green European electricity market', Public Report (WP7); Claus Huber, Reinhard Haas, Thomas Faber, Gustav Resch, John Green, John Twidell, Walter Ruijgrok, Thomas Erge; Energy Economics Group (EEG), Vienna University of Technology, Austria, 2001

Box 8.2

Renewable energy in Europe

Europe is seeing massive expansion in renewable energy development with environmental concerns being a major impetus. The following is in no way a comprehensive review, but rather is an attempt to give a feel for the pattern of development.

Denmark has been a pioneer, in part due to its decision in the 1970s not to develop nuclear power, with wind power being its most obvious commitment, along with a significant energy conservation programme. The financial support structure it operated involved a range of subsidies for operators of wind turbines and other renewable systems. Interestingly, there is widespread local ownership of the wind project – around 80 per cent are owned by local people, including many involved with local co-ops, the so-called 'wind guilds'. The wind turbines that were developed in Denmark in the 1970s and 1980s proved to be world beaters, and Denmark now dominates the world market. Domestically, it has the world's highest use of wind power. By 2002, it supplied 18 per cent of Denmark's electricity. Indeed, in some areas, at some points in the daily demand cycle, wind power was providing almost all the power. It was also a pioneer in offshore wind, with a 20-MW wind farm being constructed off the coast near Copenhagen and there are plans for several more. Unfortunately, however, following the election of a conservative administration in 2001, the various subsidy schemes supporting the development of renewables were cut back, including Denmark's ambitious wave energy programme which, as was noted in Box 7.3, was expected to repeat the success of its wind programme. But two more large, offshore wind farms were allowed to go ahead, the first, completed in 2002, being a 160-MW wind farm at Horns Rev. with eighty 2-MW turbines.

Sweden followed the Danish example, having decided to try to phase out its nuclear plants. It initially developed some large wind turbines, and allocated some $40 million to wind power between 1991 and 1996. It also has an offshore wind farm programme, and a major biofuels programme. The *Netherlands* has a significant wind programme, which initially benefited from grants of up to 40 per cent of the capital cost of wind projects and then a REFIT-type scheme. Its first offshore wind project was a 2-MW project near Medemblik. However, continuing support for renewables has been threatened by political shifts similar to those that occurred in Denmark, with the tax subsidies for renewables being cut, but in late 2002 the new Dutch government fell, after only 100 days in office, and the future path is unclear as yet.

In general, the focus in each country is shaped not just by its politics but also by its geography and the renewable resource that this defines; for example, countries with North Sea or Atlantic coasts are interested in wave energy. Thus *Norway*, which already generates the bulk of its power from hydro, has also been developing wave energy systems, as has *Portugal* and, of course, the UK and Denmark.

By contrast, most of the southern countries tend to focus on solar power. For example, the EC has supported large-scale solar thermal projects in *Italy* and active solar is widely used in *Greece*. But both countries are also interested in wind power, and *Spain* has a significant wind programme, with something of a 'wind rush' being underway.

Biofuels, like rape methyl ester, or 'bio-diesel', have also come to the fore in many European countries where there is plenty of agricultural land area, particularly in *France*, where land is set aside to grow this energy crop. *Austria* also has made extensive use of biomass for fuelling district heating networks.

Tidal power is another geographically determined technology: the UK has some excellent sites, but so far Europe's only major tidal energy project is a 240-MW barrage on the Rance estuary in Brittany in *France*, which was completed in 1967, providing something of a contrast to France's large nuclear programme. Interestingly, however, following a review of energy policy in 1995, with a new socialist–green coalition government, France seems to be making a much larger commitment to renewables. In 2001, a target of a 21 per cent contribution from renewables was set for 2010, with up to 10 GW wind capacity being planned. Renewables (including hydro and biomass) currently supply 15 per cent of France's electricity.

Germany, which was actually a relatively late entrant into the renewable energy field, has followed Denmark's example and is supporting wind power heavily, with high levels of national and local state support, via grants and market subsidies. By 2002 wind was supplying 3.5 per cent of Germany's electricity, with 12,500 wind turbines installed. Support for photovoltaic solar cells in Germany has also been very strong. Overall, by 2002, renewables were supplying 8 per cent of the country's electricity, and the aim is to go beyond the EU target in Table 8.1 and expand the renewables contribution to 20 per cent by 2010. Germany clearly sees renewables as being of major importance. It is not hard to see why – it imports over 55 per cent of its fuel and, in the late 1990s, its red–green coalition government decided to phase out nuclear power over the next two decades. Germany's policy on renewables has also been reinforced by concerns over global warming and environmental pollution. Following the UN 'Earth Summit' Conference on Environment and Development in Rio in 1993, Germany committed itself to a 25 per cent reduction in carbon dioxide emissions, and it was the first EU country to ratify the 1997 Kyoto climate change treaty.

Outside of the EU, there are smaller renewable energy programmes emerging in *Central* and *Eastern Europe*, in Poland, Hungary, Turkey and so on, some with support from the EU. Interestingly, with European Union enlargement in mind, the EU seems to see involvement with renewable energy projects as one of the criteria for membership. On cue, Turkey, having decided not to invest in nuclear power, is investing in wind power, with which it is well endowed and similar projects are emerging in other candidate countries.

Russia had done some work on wind power and there were, at one time, plans for major tidal energy schemes. However, while the energy resource is large, given the economic crisis following the break up of the USSR, there are likely to be major problems in developing it, not least getting access to the requisite funding. Certainly, given the generally very polluting and inefficient energy generation systems that exist in many of the ex-Soviet countries, support for renewable energy technology and improved energy efficiency ought to be a major priority.

US renewable energy programme

As in the UK, a high level of emphasis on cost effectiveness has been a touchstone in the approach adopted to renewables by the USA in recent years, but in the early years, the US renewable programme was very much driven by other considerations. For example, in the 1970s, following the first oil crisis, President Carter established Project Independence with the aim of obtaining 20 per cent of the USA's power from renewables by 2000. Although that may not have happened, in a country with so much material and financial wealth, the scale of renewable development has often dwarfed that elsewhere, at least until recently. That was particularly clear in the 'wind rush' that occurred in California in the late 1970s and early 1980s when thousands of machines were installed, although not all of them were well designed, with some subsequently failing.

Following the 1974 oil crisis, large-scale Federal Government R & D funding had been provided for renewable energy, rising to over $1.2 billion in 1979–80. Funding was cut back dramatically during the Reagan and Bush years, but, with the Clinton administration and the advent of concern over global warming, it subsequently grew steadily, from $119 million in 1990 to $240 million in 1992, rising to $274 million in 1994, with a $337 million programme being proposed for 1995. However, in 1996, the Republican majority in the US Congress imposed cutbacks and the election of George W. Bush in 2000 led to R & D funding being cut again, although Bush has put renewables alongside nuclear and coal as his favoured energy options, and some funding still continues to flow.

The USA has a large existing hydro capacity, but the development of the 'new' renewables has brought the total renewable generating capacity to more than 15 per cent of total US electricity generation capacity. Of the 'new' renewables (i.e. apart from hydro), various types of energy crops have been developed most extensively (biomass represented 50 per cent of the total renewable energy produced in the USA in 2002), followed by geothermal (6 per cent) and wind and solar energy (both at 1 per cent in 2002). Interestingly, according to the US Department of Energy, the hydro share in 2002 was 42 per cent of total renewables, but its contribution had fallen by 23 per cent from previous years due to the record low rain and snow falls, this providing a reminder that some renewables may be affected by climate change.

Overall, in 2002, around 53 per cent of the renewable energy produced in the USA was delivered as electricity, with there being around 97 GW of renewably sourced electricity-generating capacity, including large hydro. Of the new renewables, wind power, at 3 to 6 cents per kWh depending on location, is seen as increasingly competitive with fossil fuels, and cheaper in some locations than oil- or gas-fired power. Wind projects are consequently continuing to develop rapidly around the country, moving out from the initial California base, for example into Minnesota, Iowa, Texas, Colorado, Wyoming, Oregon and Pennsylvania. This has boosted US wind generation capacity to over 4000 MW. Some very large wind projects have been built, including a 300-MW wind farm on the Oregon/Washington border, currently the world's largest, but there are also plans for one ten times larger in South Dakota. There are also plans for a giant offshore wind farm, covering an area of 28 square miles, 4 miles off Yarmouth on the New England coast. In parallel, photovoltaic solar systems are also gaining ground, following President Clinton's commitment to a 1-m solar roof programme.

Even so, compared with the huge potential of renewables in the USA, the scale of development is still relatively limited. By 2002, around 4 per cent of US primary energy was being produced from new renewables, overtaking the 3 per cent from existing hydro. Clearly, this is only scratching the surface. Lester Brown from the Washington-based Earth Policy Institute has claimed that 'the Bush plan to add 393,000 megawatts of electricity nationwide by 2020 could be satisfied from wind alone', arguing that 'the U.S. Great Plains are the Saudi Arabia of wind power. Three wind-rich U.S. states – North Dakota, Kansas, and Texas – have enough harnessable wind to meet national electricity needs' (Brown 2001).

The support system for renewables in the USA has gone through a number of changes over the years. As indicated above, the federal government has at various times provided significant funds for R & D, while some states have provided tax concessions and other forms of subsidy designed to stimulate the market. A move has been made in some states to a target-based 'renewable portfolio standard' (RPS) system, designed to be both competitive and provide a secure market for renewables (see Box 8.3). However, the increasing level of competition in the US energy scene, and the deregulation of energy markets, has had an adverse effect. Nowhere was this more apparent than in California, where, in 2000–1, following the deregulation of the energy market, there were regular blackouts and sudden large price increases.

As in Europe, liberalisation (or 'deregulation' as it is called in the USA) is seen as a way to increase competition and reduce prices, but the

---

Box 8.3

The US Renewable Portfolio Standard system

The Renewable Portfolio Standard (RPS) is a target-based system, without fixed prices, designed to encourage the development of wind, solar, geothermal and biomass energy. RPS basically sets out net electric sales requirements of renewable energy for power producers in relation to their total energy supply, and, although prices are left to market competition, there are sometimes price caps, production credits and/or tradable renewable credit arrangements. Versions have been introduced on a state by state basis. By 2002, with California joining in, thirteen states had renewable portfolio standards of varying effectiveness. According to the American Wind Energy Association, the most successful state programme so far was signed into Texas state law in 1999 by then-Governor George W. Bush. However, in his role as US President, Bush seems to be less keen to apply the RPS nationwide. The Bush administration has argued that it was up to individual states to 'set the generation mix', pointing out that 'many states do not have renewables such as hydropower and wind'. Many retail providers would therefore not be able to meet the renewable standard; and would be required to buy renewable trading credits, which could push up the price of wholesale power. The counter argument, paralleling the view of the EC in Europe, is that energy is a traded commodity, traded across state boundaries, and there is, therefore, a need to have some harmonisation of rules. Otherwise companies could have fifty different standards to meet.

---

way this played out in California was disastrous. There seem to be as many explanations of exactly what happened, and why it happened, as there are analysts. But it seems that the key to the crisis was that a fixed consumer price ceiling was imposed. When there was an upsurge in demand, wholesale prices shot up, and some of the key power companies collapsed since they could not (initially) pass on the extra cost to consumers and could not operate with the lower profit margins. So it was an institutional and regulatory crisis rather than a technological one, although there was a technological element. Economic pressures in the electricity market, coupled with environmental controls on plant construction, had meant that there had been less investment in new capacity and in grid maintenance and development, with the result that, when electricity demand began to rise (particularly as a result of the widening use of computers and air conditioners), the system could not cope and importing power from outside of California was both difficult and expensive. It has also been argued that some of the supply companies were faced with what are rather quaintly called 'stranded assets' in the form of nuclear power plants, which could not compete in the new market but had to be kept going and paid for. It was also claimed that some of the power shortages were in fact contrived, and that some of the companies involved had manipulated the market, withholding power to boost profits. Illegal market manipulation seems to have been confirmed by the former chief trader of Enron's West Power Trading Division in Portland, Oregon, who in October 2002 agreed to plead guilty to a charge of conspiracy.

Whatever the full explanation of the cause, the result was clear. With demand exceeding supply, wholesale prices shot up (by a factor of nearly 50 by January 2001) and consumer prices shot through the ceiling. But that was not enough and a rotating, area by area, blackout programme had to be imposed.

The panic over the Californian power crisis led to calls on one hand for something akin to public control by the state, since the private sector evidently could deliver, and on the other, somewhat inexplicably given that they seemed to be part of the problem, to proposals for the construction of new nuclear plants, to meet the alleged 'energy gap'. Renewables meanwhile were somewhat disadvantaged by the highly competitive climate, but some large wind projects and the expansion of renewables generally seem likely to provide at least some respite from California's problems. California's Governor, Gray Davis, has endorsed new state RPS legislation aimed at increasing the amount of electricity produced from renewables from 12 per cent to 20 per cent by 2017. A California Consumer Power & Conservation Financing Authority was set up and has started to develop 1,800 MW of new capacity at a cost of US$2 billion, part of a programme designed to support 3,500 MW of sustainable energy capacity by 2006, including 2,400 MW of renewable energy generation capacity. The new Financing Authority can float up to $5 billion in bonds to finance projects, which would be repaid from the sale of electricity to investor-owned utilities.

The Californian crisis may lead to a more robust system developing in the future and the USA continues to be technologically innovative, with renewables clearly being well placed to play an increasing role in the US energy scene. However, with the US government refusing to ratify the Kyoto treaty and thereby, as many see it, effectively ducking its responsibilities for dealing with the USA's massive contribution to climate change, most people are looking elsewhere for sustainable solutions to our environmental problems.

Japan's renewable energy programme

Given Japan's role as a major player in the innovation field, and its pivotal role in hosting the key UN Climate Change negotiations in Kyoto in 1997, it is perhaps worth looking at how it has approached renewable energy development. The Kyoto accord (which it ratified in 2002) requires Japan to cut greenhouse gases by 6 per cent from the 1990 level in the 2008–12 period. It will clearly be a challenge to meet this target. Japan has few indigenous energy supplies and imports the bulk of its energy.

As elsewhere, interest in renewables started following the oil crises of the mid-1970s. Solar heat collectors are now very common in Japan, as is the use of geothermal heat for hot water supplies, and also for electricity generation. As noted earlier, a wave power programme has also been underway. Although there are relatively few suitable sites for wind farms on the Japanese mainland, Japanese companies like Mitsubishi have been successful at developing and exporting wind turbines. For example, there are Japanese-equipped wind farms in California and in Wales, and Japan is opening a market for its wind systems around the Pacific Rim area. In Japan itself, by 2001, 160 MW of wind capacity had been installed, and there were plans for offshore wind projects and an overall wind target of 3 GW by 2010.

There has also been significant investment in photovoltaic (PV) solar cell development: Japan has launched a '70,000 roof' PV module deployment programme. Japan evidently sees PV as a major option for the future, possibly in conjunction with hydrogen production. Japan seems keen to explore the idea of the 'hydrogen economy', that is, the production of hydrogen by electrolysis, using renewable sources of electricity, and its transmission to the point of use along gas mains, where it can either be burnt directly to provide heat or electricity or converted into electricity in a fuel cell. Hydrogen could thus become a new energy carrier, paralleling the transmission of power by electricity.

The development of renewables in Japan has been carried out under a variety of programmes, with government planning agencies working closely with private sector companies and universities. The 'Sunshine' alternative energy R & D project was started up in 1974 by MITI, the Ministry of International Trade and Industry, via its Agency of International Science and Technology. There was also a 'Futopia' wind power sub-project: 'Fu' is the Japanese word for wind. A parallel energy conservation technology programme was set up in 1978, the 'Moonlight' project.

Overall MITI has played a key role in developing renewables, with collaborative private/public sector planning arrangements helping it to map out areas for strategic investment. A New Energy and Industrial Technology Development Organisation, NEDO, was set up in 1980 to 'promote the consistent and systematic transition from oil to alternative energy sources'.

In 1993 the Sunshine and Moonlight projects were integrated within a New Sunshine programme aimed at responding to the global warming problem by cutting Japan's carbon dioxide emissions by half by 2020, with alternative energy technologies and energy conservation expected to account for one third of Japan's energy consumption. The New Sunshine programme's budget in 1994 was Yen52.8 billion and it is expected that by 2020 a cumulative total of some Yen1.55 trillion will have been allocated to it, equivalent to Yen55 billion p.a. or around £370 million p.a. (Bouda 1994).

In 2002, the Ministry of Economy, Trade and Industry (METI) announced plans to push the domestic renewable energy market, by obliging power retailers to obtain 3.2 per cent of the energy they sell from energy generated by solar, wind and other types of renewable sources by April 2003. If the proposed legislation is passed, an annual target will have to be set by each company, and the ministry will fine them up to Yen1 million if they fail to comply with the law. METI will set the aggregate targets for the use of new energy in the coming eight years as the basis for annual target calculations by each firm. Each firm will be required to report to the ministry its specific target for the coming year and results from the preceding year. As with the UK's Renewables Obligation, the firms would be able to achieve their targets either by generating new energy with their own facilities, buying electricity from these authorised new energy generators, or buying surplus from other retailers.

In addition there are plans for local ownership of some projects. For example, in response to a government-led invitation, if plans go through, a subsidiary of Tokyo Electric Power Co. is hoping to get 2,000 Tokyo residents to invest in two 1-MW wind turbines, costing Yen430 million, on reclaimed land in Tokyo Bay. Investors would receive an annual dividend of 1–3 per cent of their investment during the operating period. Their initial outlay, which is considered an investment rather than donation to environmental causes, would be repaid after about fifteen years. The electricity would be certified as 'green energy', under the government's new green power retail programme, and sold at a premium price to companies.

Clearly, Japan is not only developing a range of renewable energy technologies, it is also developing innovative implementation and support ideas.

Conclusion

The overall global renewable energy potential is considerable. A study produced for the UN Conference on Environment and Development in 1992 concluded that renewables could supply 60 per cent of the world's electricity and around 25 per cent of its heat requirements by 2025 (Johansson *et al.* 1993).

Even if very ambitious scenarios like this are discounted, the potential for renewable energy still looks very good, and, as has been indicated, commercial, strategic and environmental considerations have led to increasing interest in renewable energy in many countries, along with energy conservation and energy efficiency, as a longer term option.

Clearly, the mix of energy options used in each area of the world will depend on the local context and resources. For example, for the UK, in its Energy Review the Cabinet Office's Performance and Innovation Unit suggested a target of obtaining 20 per cent of UK electricity from renewables by 2020, with on-land wind and biomass being the main early entrants, possibly followed by PV solar and the marine renewables – offshore wind, wave and tidal current turbines. That would obviously make sense for a maritime nation with limited land area but excellent offshore energy resources and well-developed offshore energy expertise.

By contrast, countries like *China* have very large land-based renewable resources, so a different pattern of development might be expected. China's on-land renewable energy resource, leaving aside large conventional hydro, is put at the equivalent of over 400 GW, which is more than the currently installed conventional generating capacity. The new renewable options include over 90 GW of small hydro power, about 250 GW of wind, approximately 125 GW of biomass energy, about 6.7 GW of geothermal energy and an abundance of direct solar energy. The current contribution from new renewables is around 19 GW, with most of this from small hydro, accounting for around 5 per cent of total electricity. Large hydro supplies around 18 per cent. For the future, wind looks like being the biggest growth area for China. On current plans, wind is expected to expand from 500 MW as at present to 3 GW by 2005 and 5 GW by 2010. Small hydro is expected to rise to 22 GW by 2005 and

25 GW by 2010. By 2005, the total renewable capacity could be around 26 GW, rising to over 30 GW by 2010 (Hongpeng 2000).

Given its population size and rapidly growing economy, China is probably the pivotal country in terms of energy patterns in the industrialising world, and so it is good to see that in 2002 China decided to ratify the Kyoto accord. However, that will not necessarily lead to a dramatic reduction in emissions. Although China's economic modernisation and rationalisation programme has improved the efficiency of its energy and industrial systems, demand for energy is increasing rapidly. At present the bulk of China's energy is produced from coal and there are large reserves, so the use of this fuel is likely to expand. The combustion of this coal is already creating major air-quality and health problems, as well as contributing to carbon dioxide emissions. While, as noted above, there are very large renewable resources and a programme of expansion, with its gas reserves being limited, China is also keen to expand its nuclear programme.

India is in something like the same situation. It has an ambitious wind programme, having already installed over 1 GW of wind capacity and plans to expand its renewable capacity to 10 GW by 2012. However, it is also interested in nuclear expansion. So while it seems that most of Europe has made its choice and is pushing ahead with renewables rather than nuclear, some countries in other parts of the world are more ambivalent.

Within the industrialised world, the fault line seems to be whether they have signed up to the Kyoto climate change accord, and, to a lesser extent, whether they support nuclear expansion. It is not just a matter of the newly pro-nuclear but anti-Kyoto USA versus the rest, led by an eco-minded EU. For one thing, there are some conflicts within the EU. Although most EU member states seem to be anti-nuclear, there are still pro-nuclear policies emerging from within the EC, and although there are clearly several EU countries with centre–left or red–green coalition governments, which are strongly pro-renewables and anti-nuclear, several countries have shifted to the right in recent years, notably Denmark, and have changed their priorities.

Outside of Europe, there is also a mixed pattern, influenced in part by location, and by politics. *Australia* supplies much of Asia with coal and also sells uranium around the world, and its conservative government decided not to ratify the Kyoto treaty, even though the accord had been adjusted to allow Australia to actually increase its greenhouse gas emissions. Moreover, despite having a strong grass roots environmental movement, the conservative administration does not seem to be developing the country's huge renewable energy resource with much sense of urgency. Its target is to obtain 2 per cent of its power from renewables by 2010. By contrast *New Zealand* did ratify Kyoto and is pushing ahead strongly with renewables and energy efficiency, and is phasing in a carbon tax. *Canada* seems to be halfway in between, keen on renewables, also keen on nuclear, but initially at least not sure about Kyoto. Whereas, as we have seen, Japan seems to be keen on all three of these elements in the policy mix.

Within the newly industrialising world, as we have seen, China and India, like many other rapidly industrialising countries, are still at an energy crossroads. They are keen to develop renewables and worried about climate change, which could well impact on them more than the other countries, but they are also keen on nuclear power.

There are clearly many choices to be made and many difficult technical and economic problems to overcome, as well as some potential for conflicts, not least over nuclear power. The World Summit on Sustainable Development, held in Johannesburg in 2002, failed to resolve this issue, calling for a 'substantial increase' in the global use of renewables, but not commenting directly on nuclear power. Leaving that issue aside, there is nevertheless something of an emerging consensus that, as the UN/World Energy Council's 'World energy assessment' report, published in 2000, put it 'there are no fundamental technological, economic or resource limits constraining the world from enjoying the benefits of both high levels of energy services and a better environment', although, a little more cautiously, it added 'A prosperous, equitable and environmentally sustainable world is within our reach, but only if governments adopt new policies to encourage the delivery of energy services in cleaner and more efficient ways' (WEA 2000).

The next chapter looks at some of the strategic issues and options that may lie ahead if sustainable energy technologies are to be widely adopted. While there may not be any fundamental technological problems in developing renewables, moving to a fully sustainable world energy system will require some major changes in the way we generate and use energy.

Summary points

- Renewable energy is being developed around the world, with the advanced countries taking a lead, notably the USA, Germany and Japan.

- The UK has a large renewable resource but is only developing it relatively slowly.

- Environmental concerns have been a major stimulant for these developments but equally there is a growing awareness of the commercial potential of renewable energy systems.

- The potential of renewables looks very significant, but there are a range of strategic issues which will have to be addressed if this potential is to be developed fully.

Further reading

Inevitably given the pace of technological change and policy shifts the outline of renewable energy development given above will date. Useful, detailed studies of global renewable energy potentials, along with discussions of all the other energy options, are included in the 'World energy assessment' report that was produced by the World Energy Council and the UN in 2000.

To keep up to date on what is happening in the renewable energy industries around the world you could consult the annual *World Directory of Renewable Energy Suppliers and Services* (James and James, London), which, in addition to listings of suppliers, includes a series of overview papers reporting on the latest developments in technology and policy worldwide. Parts of it are available on the World Wide Web at http://www.jxj.com/index.html.

The international renewable energy journal *CADDET*, produced in liaison with the International Energy Agency, provides useful reports of developments around the world. It is available on the World Wide Web at http://www.caddet.co.uk/.

The contacts list in Appendix II includes some other useful sources of information on renewable energy developments around the world.

9 Sustainable energy strategy

- Security of supply
- Compensating for intermittency
- Conservation versus renewables
- Scale and pace of deployment

A shift to a sustainable energy system would require a number of technical, economic and strategic issues to be addressed, not least the fact that some renewable energy sources are intermittent. This chapter looks at some of the emerging issues, including the problem of intermittency, the debate over whether to focus on energy conservation or on new energy supply technologies, and the question of the pace and timescale required for the development of sustainable energy systems.

Moving to a sustainable energy system

Along with energy conservation and the more efficient use of the world's remaining fossil fuels, renewables look like they can help the world move towards a sustainable energy supply and demand system. Of course, whether complete sustainability can be achieved in this way remains unclear: it depends on a range of issues, not least the level of economic growth that is being aimed at around the world. Wider issues like this are explored in Part 3 of this book. But clearly the success or otherwise of any attempt to move toward sustainability will depend on the degree to which novel technologies in the renewable energy and energy conservation field can be developed and deployed.

The emphasis in this chapter is on renewable energy, but this is not to suggest that the energy conservation side is less important: they represent two sides of the same coin. There is no point in pushing ahead with new energy supply technologies unless energy demand is held down and, ideally, reduced.

Although, given the dominance of supply side thinking, energy conservation is still often underfunded, some progress has been made in recent years. This is in part because, quite apart from any environmental concerns, investing in the more efficient use of energy makes sense in economic terms. The technologies involved may not always be as exciting as the new renewable energy technologies, but they are important, and for the foreseeable future this is where a lot of the practical day-to-day progress will be made.

In addition to the various technical means of energy saving, there is also a range of planning techniques which are likely to become increasingly relevant as part of a process of managing energy demand, such as least-cost planning and integrated resource planning. At the same time, much is happening on the supply side and later in this chapter we shall be discussing what might be the most appropriate balance between supply side and demand side.

Security of energy supply

On the supply side, a key element in any sustainable energy system must be that it is robust and can continue to provide power regardless of changed circumstances. Clearly, whatever the energy system, it is important to have secure energy supplies.

Security of supply can be ensured by having a diverse range of fuels and technologies and by selecting fuels that are not likely to be prone to interruptions in supply. Following the oil crisis in 1973–4, this was why many countries tried to diversify away from oil. There were also concerns about the impacts of industrial disputes on fuel supplies, for example, following the various confrontations between workers and governments over the future of the UK coal-mining industry.

More recently it has been argued that since renewable energy systems involve a range of scales of technology, using a diverse range of locally available indigenous sources, they are likely to offer a more reliable basis for secure energy supplies than systems that rely on imported fuels. Natural renewable energy flows are unlikely to be interrupted by human intervention, so there is no need to stockpile fuel. Terrorist attacks seem somewhat less credible with a national network of wind turbines than, say, with a handful of nuclear plants or offshore oil rigs.

However, if renewables are to be widely used, the problem of the *intermittency* of some of the sources has to be faced – the sun does not always shine, the winds do not always blow and waves do not always rise. Not all renewable sources are intermittent. The use of biofuels, hydroelectric power and geothermal energy can provide 'firm' (i.e. continuous) power and tidal energy is very predictable, but the availability of solar, wind and wave energy is not so predictable.

Fortunately, some of the basic annual weather cycles link up well with human energy requirements. For example, in many locations, peak daytime air-conditioning loads in summer link well within peak power availability from solar PV systems. At the other extreme, in much of the world it is windy during the winter so wind and wave power is at a peak when electricity is most needed for heating. Indeed there is a further correlation. Wind produces a chilling effect on buildings and conventional power systems have to take this into account by having extra generating capacity ready for windy days. In the UK this amounts to around 1 GW. However, if we have 1 GW of wind capacity on the system, its output will be positively correlated to some degree to the increased demand from the wind chill effect.

The shorter term variations in renewable energy inputs are more of a problem, since there can also be negative correlations between availability of renewable energy like wind and energy demand. For example, the wind does not blow continuously, so some wind turbines will be unproductive at any particular time, and this may coincide with cold weather. On this basis some critics have argued that intermittent renewables like wind are unreliable energy sources (Laughton 2002).

Certainly, at first glance, it would seem to be reasonable to think that energy generation that depended on something as variable as the weather system would be unreliable compared with generation technology that could supply power continuously. Perhaps surprisingly though, in practice, up to a certain level, this intermittency need not be a major operational problem, if the electricity from these devices is fed into an integrated national power grid network, since this can 'even out' and 'buffer' the local variations in energy inputs. Individual wind turbines may be becalmed occasionally, but in a country like the UK it is usually windy somewhere, so other wind plants will be delivering power to the grid, a situation that will be further improved as wind farms are built at increasing distances offshore. Moreover, it is not the case that conventional power plants are 100 per cent reliable, and to deal with variations in the power available from conventional generators, and also with the variations in energy demand, the grid system is usually operated with some conventional plant on standby – it is kept running at low power as a so-called 'spinning reserve'. This can be used help compensate for any variations in supply from renewable sources, depending on the amount of intermittent renewables involved.

In practice, so far, intermittency has not proved to be a problem. By 2002, the UK had around 3 per cent of electricity coming from renewable sources, but the variations in supplies from the 500 MW of wind farm capacity could not be detected by the grid controllers – the variations were lost in the 'noise'. This is hardly surprising, since the grid deals with much larger disruptions and disturbances, due, for example, to the occasional sudden loss of power from a large conventional power plant (up to 1 GW or more), lightning strikes on the supergrid (2,000 MW), temporary failure of the cross-channel link (1,000 MW) and even the start up and shut down of Eurostar trains (5 MW), as well as the other more regular and larger variations in demand patterns, driven by consumer behaviour. However, if the total contribution from wind power was more than about 10–15 per cent of the total power being supplied to the national grid, the variable nature of the inputs to the grid would begin to have more impact (Milborrow 2001).

So far there do not seem to have been significant problems with the Danish system, which by 2002 was operating with an 18 per cent contribution from wind. Indeed, at times, in some areas of the country, during periods of low demand (e.g. at night), most of the conventional plant was run down, leaving wind power as the main input, and at some other times, when there was more wind power than was needed, the excess energy has been used to provide heat to boost the accumulator heat store units associated with combined heat and power plants, of which Denmark has many (Toke 2002).

Of course, this does not mean that there could not be times when weather conditions would reduce the net wind contribution considerably, and this would become more important as the percentage contribution from intermittent sources increased. So at some point there could be a need for some extra standby plant, or for short-term energy storage facilities, to maintain full grid security.

For the moment, there is no shortage of standby plant – most power systems operate with substantial 'plant margins' and with excess capacity over and above what would be needed to meet peak winter demand. In 2002, the UK excess capacity was around 25 per cent. Maintaining and operating standby plant does cost money, although, at present, it is cheaper than providing energy storage. But, however it is achieved, dealing with intermittency need not be prohibitively expensive. In its UK Energy Review, published in 2002, the Cabinet Office's Performance and Innovation Unit (PIU) concluded that the extra cost of dealing with intermittent sources like wind would be only 0.1p per kWh up to a 10 per cent national contribution and 0.2p for 20 per cent. Even with a 45 per cent national contribution from intermittent sources, the extra cost would they concluded only be around 0.3p per kWh (PIU 2002).

In addition to the extra cost, the provision of back-up power would of course involve extra emissions, although some of the plants could use stored renewable fuel, such as biomass. But, given that the periods for which they would be used would not be long, the use of fast start-up gas turbines might be condoned, despite their carbon emissions. In some circumstances, energy storage could also be a viable option. Short-term storage could be provided by a variety of small- to medium-scale mechanical and electric systems (e.g. compressed air, flywheels, batteries) or by large-scale pumped storage schemes, although these techniques are all to varying degrees expensive. However, new electricity storage techniques are also emerging, such as the Regenesys system developed in the UK, which uses a Redox chemical reaction. A 15-MW prototype unit has been installed alongside a gas-fired power station near Cambridge, designed as a grid back-up system, at a cost of around £20 million, with subsequent versions expected to be cheaper.

As we have seen, for the moment, the existing power system can in effect act as a back-up to intermittent inputs from wind, with the need for extra back-up plant or energy storage being limited. Moreover, the power system is in any case changing; for example, the trend to smaller conventional power plants is making load following easier, so that the impact of demand peaks can be less pronounced. At the same time, the adoption in key sectors of improved energy efficiency measures and of demand management techniques can reduce peak loads, making it easier for the power system to cope with larger contributions from intermittent renewables. For example, in the domestic sector, to deal with peak demand, interactive load management systems, such as 'smart plugs', can allow consumers to avoid periods of high power prices by switching off selected devices like freezers – which can happily run for a few hours without power. In addition, there is an increased reliance on power imports from neighbouring countries, which can help compensate for local intermittency in supply as well as local demand peaks. That seems to be one reason why the Danish system is operating successfully with 18 per cent of wind on the grid – it can import balancing power from Scandinavia (e.g. Norwegian hydro) when and if it is required.

Finally, we have focused so far just on wind. If other renewables, like wave power, are also feeding into the grid, the overall problem of intermittency may be lessened since the timing of the variations in power availability differ. For example, wave energy is in effect stored wind energy, which will be available long after the wind has dropped. In addition, energy from tidal flows, although cyclic, is unrelated to wind, and peak solar energy availability similarly depends on weather patterns not directly linked to wind patterns. With all these inputs feeding into the grid, the impact of variations in inputs from any one source could be reduced. One early study suggested that, assuming the provision of storage capacity, demand management and trading of power with neighbouring systems, a system using a range of intermittent renewables could be developed that was technically viable even if the contribution from variable renewables reached over 50 per cent of the total energy used (Grubb 1991).

This 50 per cent estimate may be on the optimistic side, especially since it ignores the cost. However, although there are still disagreements on the exact scale of the operational penalties and costs, it seems that intermittency is likely to be a technical integration problem, not a fundamental flaw, at least for moderate levels of variable inputs to a national grid system, perhaps even up to around 30–40 per cent of the total, from a range of intermittent renewables.

The hydrogen economy

Fortunately, if we want to go beyond moderate levels of renewable contribution, there is a new, potentially cheap, longer term energy option opening up, which can help to balance the energy system – the use of hydrogen gas. Hydrogen can be generated, when power is available from intermittent renewables, by the electrolysis of water and then stored for later use. Electrolysis provides a reasonably efficient way of using electricity to split water into hydrogen and oxygen: it has an energy conversion efficiency of up to 80 per cent. The hydrogen gas could be stored or transmitted down conventional gas grids to where it was needed. In the interim, while natural gas is still available, it could be mixed with hydrogen. In the longer term, there might be a shift in emphasis from electricity transmission to hydrogen transmission. Gas distribution is much more efficient than electricity transmission and gas can be stored more easily than electricity. When burned, hydrogen produces only water as a byproduct. Although care has to be taken with this fuel, especially if stored as a liquid in cryogenic form, its use should not require much more in the way of safety provisions than we are used to with petrol, which after all is a very volatile fuel. In some ways hydrogen is safer than many conventional fuels. Being lighter than air, hydrogen gas rises, so any leaks would vent out of houses rather than collecting on the floor and building up until they reached a source of ignition, as can happen with conventional natural gas.

A switchover to what has been called 'the hydrogen economy' has been seen as a possible alternative both to oil and gas distribution and to electricity transmission, with giant hydrogen gas grids around the world and hydrogen also being tanked around in liquid form in ships (Rifkin 2002). Indeed this is already being done with hydrogen from Canada's hydro plants being shipped to Sweden. In the longer term countries in the Middle East might install large arrays of photovoltaic solar cells in desert areas to generate hydrogen to supply the less sunny areas of the world with a new clean fuel. Hydrogen can be used as a clean fuel for vehicles or can be converted into electricity in a fuel cell. Fuel cell-powered vehicles are already emerging, and fuel cells are also increasingly being used for stationary power generation. Fuel cells can be highly efficient energy converters, with energy conversion efficiencies of up to 80 per cent, especially if the heat produced is also used. A recent study suggested that by 2012 there could be 16,000 MW of stationary fuel cell capacity in use. It could thus be that in future there will be a global system based on renewable hydrogen, with electricity only being generated locally where and when needed.

Would that amount to just a very large-scale technical fix? That rather depends on how the system was designed, developed and run – and in whose interests. Technical possibilities like this seem likely to open up a whole new range of strategic debates concerning how systems should be developed in future.

Strategic choices

Perhaps the most urgent strategic issue is the question of the *pace* and *timescale* of renewable energy development.

Given that natural gas is cheap and reasonably clean, it might be thought to be premature, and economically unrealistic, to try to push ahead rapidly with the installation of renewable energy technology. It could be argued that, instead, the breathing space offered by gas should be used to research and develop the renewables more incrementally, prior to deployment on a wide scale.

There is certainly a social case for an incremental approach: it would give time for the social processes of evaluating and negotiating trade-offs between local disbenefits and global benefits to be carried out in an open and participative way. Continuing to rely mainly on gas, while reserves last, could also be seen as less environmentally problematic if underground sequestration of carbon dioxide (in old oil and gas wells, or in aquifers) proves to be viable on a significant scale – and a reliable way to store carbon. Some would also see coal as coming back into the picture on this basis. At the same time, even if these sequestration options are available as interim measures, it is unlikely that all the power plants, old and new, around the world, will be able to take advantage of them, so that carbon emissions will continue to be a problem. Moreover, a significant amount of energy would be needed to capture and sequester carbon, and this would not only add to the cost and emissions, but would also increase the rate of use of the remaining fossil fuel reserves.

So there would seem to be an environmental case for pressing ahead with renewables quickly, so as to move up the learning curve, in terms of both technical development and social deployment. Of course that does not necessarily mean that a rapid, radical, 'crash' programme is needed: the vision should perhaps be radical, and there should be a sense of urgency, but there are still merits in a careful, incremental approach even to the development of radical new technology.

Conservation versus renewables?

Another short- to medium-term strategic issue concerns the relative merits of renewables and energy conservation. Since, at least up to a certain point, energy conservation is usually cheaper and quicker than developing new generation capacity, what should be the investment priorities in the short to medium term? The basic strategic criteria outlined earlier suggest that conservation should be preferred over generation; that is, wherever possible investment should be targeted on techniques that cut demand, rather than on simply increasing supply.

However, as with some renewables, conservation initiatives can face implementation problems. While most of the initial wave of energy efficiency measures may be relatively easy and cheap to deploy, some will require actions and purchases by individual consumers, and stimulating this type of initiative in a coherent and effective way is sometimes difficult. There is also a range of other implementation and uptake problems, not least institutional resistance to change, which may limit the extent to which the large, theoretically possible, energy savings can actually be achieved in practice in the short to medium term. We shall be looking at some of these problems in Part 3.

It is to be hoped that these problems can be resolved gradually and conservation can play its vital role, along with renewables, in helping to move towards a sustainable energy supply and demand system. It is not a matter of choosing between renewables or efficiency: in general, they are complementary. Both are needed and both need to be deployed as rapidly as possible. However, there may be some specific tactical conflicts. For example, although in general conservation projects ought to be a prerequisite, in some circumstances it may be that renewable energy projects, especially smaller scale 'modular' projects, will prove to be more effective and easier to deploy. There may also be 'technical' conflicts; for example, in a well-insulated building equipped with energy-efficient end-use devices, the level and pattern demand for energy is changed, so that the use of domestic-scale solar power may be less cost effective. And if there was widespread adoption of energy-efficiency measures in all sectors, then the overall pattern of demand would also change, and that would influence the choice of renewables to some extent. Clearly, what is needed is an optimal, integrated approach and an appropriate mix between conservation and renewables.

However, there remain some more general strategic uncertainties as to the correct mix and over the right overall strategic emphasis. For example, it is sometimes argued that, since the developed countries currently are the most wasteful, they should give energy efficiency a high priority, at least initially. Giving evidence to a UK Government review on European Energy Policy, Greenpeace International proposed a 3:1 ratio, calling for a Europe-wide commitment to a 1 per cent p.a. increase in renewable energy use and a 3 per cent p.a. reduction in energy use via investment in energy efficiency (Greenpeace 1995).

An alternative view is that the developed countries should take the lead and give the development of renewable energy generation techniques the highest priority, both for their own use and, equally if not more importantly, for use by developing countries. Otherwise the newly industrialising countries will make use of fossil fuel sources. On this view, since the advanced countries at present have the expertise, they should develop and then transfer renewable energy techniques for the developing world to use.

A hegemonic battle?

Clearly, the strategic perception influencing some of the views expounded above is that, regardless of the precise balance between them, renewable energy technology and conservation are alternatives to fossil fuels and that these alternatives must gradually, or perhaps even quite rapidly, displace and replace fossil fuels, as well as replacing nuclear power.

In strategic terms, the suggestion is that, in order to attain sustainability, it will be necessary to weaken the dominance and hegemony of conventional power supply technology, that is, their political, economic and institutional predominance must be challenged. Thus renewables must challenge conventional supply options on their own ground, i.e. in terms of generation, while conservation must cut demand and open up crucial energy end-use issues, and both can, to varying degrees, challenge the status quo of fossil and nuclear generation.

In part, this confrontational view stems from the conviction that many environmentalists share that the powerful fossil fuel industries will inevitably seek to marginalise rivals like renewables and, unless challenged, will continue to enjoy the strong backing of governments, to the detriment of renewables and conservation. Certainly, that has often been the experience in the past; for example, between 1990 and 1995 the EU allocated £2,124 million in subsidies for fossil fuel compared with £528 million for renewables (Greenpeace 1997). The imbalance has been even clearer in the case of nuclear power, which has enjoyed spectacular levels of government subsidy in many countries. For example, in 1999, the main industrial countries allocated $3,205 million to R & D on nuclear fission, £1,003 million on nuclear fusion, but only $536 million on renewables (IEA 2001b). Given that there is inevitably a limited amount of money available to support energy developments, new entrants will inevitably have to challenge the incumbents for access to funding and political support.

Not everyone sees it in quite such confrontational terms. For example, Shell International has argued that there is no need for direct conflict between old and new technologies and that, in effect, there is room for all the various means of providing energy to expand, at least for a while, since overall global energy demand will increase in the decades ahead.

In Shell's 'sustained growth' global energy scenario (Figure 9.1), a summary of which was published in 1995, the use of fossil fuel, hydro and nuclear power are seen as continuing to expand until 2020 to 2030, at which point a plateau is reached. Subsequently, there are some reorderings within this plateau pattern: oil and then coal begin to decline, but gas stabilises, while nuclear and hydro still continue to expand gradually. At the same time, the 'new' renewables begin to lift off dramatically, supplying about 50 per cent of world energy by around 2060. However, overall energy demand would rise by about a factor of three, so there is relatively little substitution of new renewables for the total fossil, hydro and nuclear fuel contribution, even by 2060 (Shell 1995).

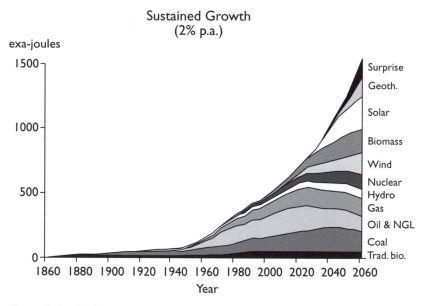

Figure 9.1 *Shell's 'sustained growth' scenario*
(*Source:* Shell International, 1995)

In fact, far from substituting for the conventional energy sources, Shell seems to see the new renewables as something of an optional 'add on'. This becomes clear in their dematerialisation scenario (Figure 9.2), in which demand is further contained by a structural shift to a more

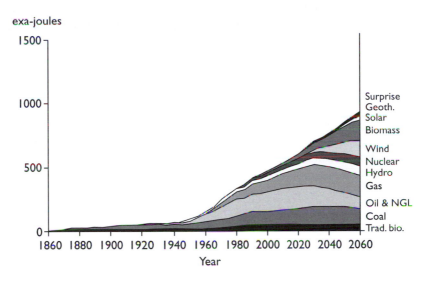

Figure 9.2 *Shell's 'dematerialisation' scenario*
(*Source:* Shell International, 1995)

energy-efficient world economy based on information technology and the use of less energy-intensive materials. The result was that demand would only double by 2060 and fossil and nuclear fuels would supply the bulk of the energy needed. The lift-off point for the bulk of the renewables would be postponed until 2040. Even then, the growth of renewables is relatively slow, and, as the Shell report puts it, the 'second wave of renewables is not needed until 2060' (Shell 1995: 15).

Shell's views on renewables seem to have changed since 1995, although the underlying message has not. In their new scenarios, produced in 2001, Shell seem even less concerned to push ahead rapidly with renewables, possibly because their overall estimate of global energy demand in their new 'spirit of the coming age' scenario in less than in the earlier 'sustained growth' scenario. Indeed, they seem to want to back away from what they call the 'dynamics as usual' scenario, in which renewables including hydro expand to around a 33 per cent primary energy contribution by 2050. They seem to be much more interested in devising a strategy for continued expansion of the use of key fossil fuels, some of these hydrocarbons (including coal) being used to produce hydrogen gas, with carbon dioxide sequestration (i.e. storage) to avoid the climate change impact. This is certainly an ambitious concept. No doubt the technology can be improved, but currently hydrogen production from coal looks very expensive, and carbon dioxide sequestration has yet to be proven as a viable option on a large scale, technically, economically or environmentally. By contrast, the generation of hydrogen using electricity produced by renewables would seem to be somewhat more straightforward, with there being no need for sequestration. However, on the basis of their 'hydrogen from hydrocarbons' approach, in their new 'Spirit of the Coming Age' scenario (see Figure 9.3), Shell sees new renewables as only needing to expand to make a 22 per cent primary energy contribution by 2050. Even with hydro included, the total renewables contribution is still only around 27 per cent. Instead the future lies with hydrogen (Shell 2001).

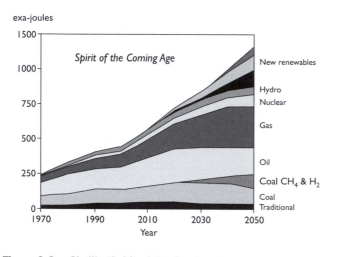

Figure 9.3   *Shell's 'Spirit of the Coming Age' scenario*
(*Source:* Shell International, 2001)

It is interesting to see a company like Shell considering a conversion from being a hydrocarbon company to a hydrogen company, much as British Petroleum has been presenting itself as moving 'Beyond Petroleum', even if in reality what they seem to be talking about is using hydrocarbons to produce hydrogen. In recent years both have invested quite heavily in renewables. In 1997 both Shell and BP said that they would be spending around $500 million on renewables. BP, who had broken ranks, like Shell, with some of the other main oil companies to admit that climate change was something they had to deal with, has focused mainly on PV solar, while Shell has branched out into wind and biofuels. Subsequently, in 2001, Shell expanded its renewable energy funding by a further $500 million. Even so these are fairly small sums compared with the huge and increasing levels of spending on fossil fuels. For example, BP's capital spending on exploration and production has risen to an average of $8 billion a year, on refining and marketing to $2.8 billion and on petrochemicals to $2 billion. Clearly, while these companies are interested in renewables, they still understandably see fossil fuels as continuing to be their core business for some time, with hydrogen production evidently being one long-term option.

Certainly, it will take time to develop renewables fully, so, in the interim, there will be a need for fossil fuels to avoid energy shortages, possibly with carbon sequestration to limit emissions, and for conservation to try to keep demand down and reduce the impact of conventional generation. But perhaps a more positive rendition of this strategy would be to portray conservation and fossil fuel technologies – such as natural gas CCGT, clean coal combustion and the conversion of fossil fuels to hydrogen backed up by carbon sequestration – as providing a 'bridge' to the solar future, rather that seeing them as ways to avoid or postpone the need for renewables.

On this basis, and reflecting the need both for 'vision' and a strategy for responding to global warming, we might tentatively add an extra, and more aggressive, strategic criteria to the list introduced in Chapter 3: *while continuing to invest in conservation and energy efficiency, and (possibly) making use of carbon sequestration, get on with developing sustainable energy generation technology as a matter of urgency.*

This does not necessarily imply a 'crash' programme of renewable energy development, and certainly not one at the expense of conservation. But it does imply that renewable energy R & D should be significantly expanded and rates of deployment increased as far as practicable, in order to ensure that a transition to sustainability is not delayed by vested interests in the existing energy system. Clearly, this is a matter of interpretation and political judgement, but it is a view that is becoming increasingly influential amongst lobbyists and in political circles (Scheer 2002).

For the immediate future the key requirement would seem to be to get moving on the development and deployment of the new sustainable technologies, and to start the long process of learning how to respond to the various implementation problems that will inevitably emerge. It is to these problems that we now turn – in the next part of this book.

Summary points

- Sustainable energy technologies may increase security of supply, for example, by reducing reliance on imported fuel.

- The intermittency of renewable energy sources can be compensated for by the use of integrated grids and also possibly by switching to a hydrogen economy.

- There are strategic debates concerning the correct balance between conservation and renewables, and the correct pace and scale of deployment generally.

- There may be a need to move quickly to develop and deploy sustainable energy systems, but there are likely to be a range of implementation problems.

Further reading

A fairly positive view of the strategic requirements for renewable energy development is taken in *World Energy Assessment*, produced by the UN/World Energy Council in 2000.

The Shell International studies, mentioned in the text, can be obtained from the Global Business Environment, Corporate Centre – Planning, Environmental and External Affairs, Shell Centre, London SE1 7NA; pxg@shell.com. See also http://www.shell.com.

A more radical viewpoint is adopted by the President of the Eurosolar lobbying group Hermann Scheer, who is a member of the German Parliament, in his forceful *Solar Manifesto* (James and James, London, 1995) and in his follow-up book *The Solar Economy* (Earthscan, London, 2002).

For a good review of the prospects for the UK see Paul Ekins' 'The UK's transition to a low carbon economy' for the Forum for the Future.

 **Problems of implementation**

Sustainable energy technologies are being developed and introduced worldwide. However, there can be obstacles to this process. Part 3 looks at some of the technical, economic, strategic and environmental limitations and problems facing the development and implementation of sustainable energy options. It focuses in particular on the institutional problems of obtaining support for the development of renewable energy technology and on the problems of winning acceptance from the public for the use of such novel technologies. Although renewable energy technologies are relatively benign in environmental terms compared with conventional energy technology, they may still have local environmental impacts and adverse public reactions can be an obstacle to implementation.

⬛10 Getting started: institutional obstacles

- Research and development problems
- Institutional resistance to change
- The need for 'follow through'

The development of new technologies is never easy. This chapter looks at the difficulties experienced by renewable energy technologies in trying to get funding for research, using the UK wave power programme as an example. It also looks at the way in which wind power has been developed around the world, highlighting the different research and development strategies that were adopted. Finally, it looks at the technical problems experienced by the UK geothermal energy programme, highlighting how prone novel projects can be to initial failure.

Obstacles to sustainable energy development

New technologies face a whole range of technical, economic and institutional hurdles as they try to get started on the long process of development and deployment. These constraints and hurdles have been particularly apparent in the case of sustainable energy technology and they may well shape the extent to which these new energy technologies can play a role in supporting a move towards a sustainable energy future – or at least they may influence the pace at which this transition occurs.

The focus in this chapter is on the early stage of the research and development (R & D) process, and, as a consequence, the emphasis is on renewables, since, in the main, that is where most of the R & D effort is needed. Most energy conservation techniques are relatively well established, the main problem is to get support for deployment. We shall be looking at deployment and implementation problems facing both conservation and renewables in the next chapter.

Here we are more concerned with the stage of getting started and in particular with the problem that researchers have experienced in obtaining initial financial support for research on renewables from governments. As will be seen, this has not always been easy, and even when it is forthcoming, the resultant projects may not always fare well and support may not be continued.

Getting started

Most new technologies need initial research and development support to get established, and obtaining access to this represents a major institutional hurdle, especially since governments are the most probable source of initial support. Inevitably, there are disagreements about which projects should be funded and over how projects should be developed. Certainly, it is hard to 'pick winners' when technologies are at an early stage of development. In addition, there may be resistance to new developments from those with vested interests in the existing range of technologies, and a lack of commitment from decision-makers to pressing ahead with what may seem like risky and long-term development programmes. In effect, we are facing the reality of the sort of battle for hegemony that we mentioned in Chapter 9.

Rather than try to analyse the institutional problems facing renewables in the abstract, it may be helpful to focus on some specific examples. We now look, via a series of mini case studies, at the problems that have faced a range of renewable energy projects. We start with a brief review of the way wave power was treated in the UK, this being one of the more celebrated, or perhaps notorious, examples of lack of follow through.

The UK wave energy programme

As was noted in Chapter 8, in 1974 the Labour Government launched a programme of renewable energy research and development. This programme, as continued under successive UK governments, although hardly generous, allowed a range of technologies to be developed and/or assessed.

Wave power was initially seen as a front runner. Glyn England, then Chairman of the Central Electricity Generating Board suggested in 1978 that it might offer enough power 'to supply the whole of Britain with electricity at the present rate of consumption' (England 1978: 272).

When a Conservative government came to power in 1979, it was evidently also supportive. John Moore, then Energy Minister, commented in September 1980 that 'whatever other problems wave energy researchers may face, lack of Government support will not be among them' (Moore 1980). The device teams in universities and elsewhere worked enthusiastically: some scale prototypes were tested in open water.

However, by 1982, views had changed, and wave R & D was cut, following assessment of all the renewables (apart from tidal, which was assessed separately) by the government's Advisory Committee on Research and Development (ACORD). The government had asked ACORD to assess the renewable options, as part of its regular review process, in the light of the expected cut-backs in overall renewable energy R & D funding (from £14 million to £11–12 million) and budgetary expenditure.

David Mellor, then Secretary of State for Energy, later commented that it had not proved to 'be possible to insulate this area from the savings the Government are making in public expenditure' (Mellor 1982).

The justification for cutting back on wave power was economic. Generation costs of 20p per kWh or more were mentioned, although the Energy Technology Support Unit, which reviewed wave power and the other renewables for ACORD, put the likely cost of wave power generation at 4–12p per kWh. Even so, it concluded that 'wave power is likely to be economic only in those futures more favourable to renewable energy technology' (Energy Technology Support Unit 1982: 14).

The full story of the ACORD review and subsequent events is complex and opens up some interesting general issues, for example, concerning the way technological innovation is handled (Elliott 1995). But the basic strategic issues are relatively clear. The approach adopted in the UK government's initial programme on wave power, from around 1976 onwards, was to establish a 2-GW 'reference design' target, and ask the wave power device teams to develop proposals for devices on that scale. Given that only very small tank-tested models were available at this stage, it was obviously quite a jump to try to come up with designs for full-scale systems, which might have 20,000 tonnes displacement or more.

Subsequently, there was some criticism of this approach as premature – asking too much too soon of an embryo technology. One commentator likened it to 'designing a super tanker whilst at the same time developing the principles of naval architecture' (Flood 1991: 44).

Small is beautiful?

It has been argued by some innovation theorists that large-scale projects like this are inevitably inflexible and do not allow an opportunity for piecemeal adaptation, incremental development, feedback and learning from mistakes. Thus, in a 1993 study of the UK wave power programme, Audley Genus argued that the 2-GW wave system design target reflected the UK energy establishments obsession with large-scale units, which he sees as having 'all the hallmarks of inflexible technology'. Thus, in the case of the wave programme, 'lead times appear to be long, capital intensity high, unit size large' and 'enormous investments of time, capital and other resources would have to have been made before any learning about actual performance and improvements of these systems could be realised' (Genus 1993: 141).

This in turn made it very hard to come up with sensible cost estimates and even harder to make sensible decisions about the future of the technology. Stephen Salter, the inventor of the 'duck' wave power system, commented that it was like trying 'to decide our aviation policy on the data available in 1910' (Salter 1981a).

Beyond that there have been allegations that errors were made in ACORD's assessment, and even that some of the more favourable assessments, by outside consultants, were suppressed. A fierce debate ensued on the way wave power had been treated. Some critics suggested that there had been a pro-nuclear bias, and there were concerns expressed about the location of ETSU at the Atomic Energy Authority's Research establishment at Harwell. For its part, ETSU strenuously protested its independence. Other critics claimed that, at the very least, the technology had been assessed at too early a stage in its development. As the all-party House of Commons Select Committee on Energy put it in 1984, the suspicion was that wave energy 'was effectively withdrawn before the race began' (Select Committee 1984: xxxii).

Certainly, there was room for disagreement amongst the experts: they faced the problem of trying to cost a range of very novel systems. As a spokesman for the Department of Energy was to tell a subsequent session of the Select Committee, there was 'definitely scope for different judgements at the early stage of the development of a device' (Select Committee 1992: 125).

There was considerable pressure from the Select Committee and from renewable energy lobbyists for a reassessment and in 1989 a new review by ETSU was set in motion. However, when the results eventually emerged in November 1992, the conclusions were similar: deep-sea wave energy was still not seen as economic and in-shore/on-shore wave energy systems were not viewed much more favourably (Thorpe 1992).

Of course it could be argued that, since very little new work had been done in the ten years since the 1982 decision to cut R & D support, this conclusion might not be surprising. Nevertheless, the result was that wave power remained in the 'long shot category' as far as the UK government was concerned and this continued to be its position for some time. Although there were some signs of industrial interest in smaller scale in/on-shore wave power, the 1982 decision certainly put a halt to work on deep-sea wave power.

It was not until a change of government, in 1998, that the wave power issue was revisited seriously. Although not much new work had been done on wave energy in the UK in the meantime, the political climate had clearly changed, in part because of growing concerns about climate change. A series of reassessments of wave energy was carried out, initially as part of the UK Technology Foresight programme, culminating in March 2001 with an admission by the Department of Trade and Industry that the decision in 1992 to abandon wave energy was wrong: 'The decision was taken in the light of the best independent advice available. With the benefit of hindsight, that decision to end the programme was clearly a mistake' (Ross 2002: 34).

Clearly, the fortunes of wave energy were changing, and the same happened in relation to tidal power, the favoured tidal technology now involving the extraction of energy via free-standing turbines mounted in the flow, rather than by large expensive and environmentally invasive tidal barrages.

In May 2001 the House of Commons Science and Technology Select Committee commented that 'Given the UK's abundant natural wave and tidal resource, it is extremely regrettable and surprising that the development of wave and tidal energy technologies has received so little support from the Government' (Select Committee 2001: iv).

Subsequently, ministers have increasingly been keen to be seen as pro wave and tidal current projects, even though the level of funding that has emerged may not have reflected the rhetoric – so far only around £6 million has been allocated to new wave and tidal current projects.

Of course, the validity of these reassessments has still to be demonstrated. But in general, rather than trying to 'pick winners' at an early stage, it seems wiser to allow a range of developments to proceed, especially since the level of expenditure at the R & D stage is relatively small. Selection becomes more important at the next, more expensive stage in the innovation process, when moving on to the full-scale development and commercial deployment. Some renewables have reached that point, and market assessments are therefore beginning to be made, but most are further back in the process.

Developing technology

The wave power story seems to indicate that new technologies face major problems in getting accepted: there can be institutional biases and a lack of vision. Thus, commenting on the way the wave energy programme had been assessed in 1982, the House of Commons Select Committee on Energy noted that there was no evidence that the Department of Energy had 'ever assessed the much larger expenditures on the fast breeder reactor and fusion research against similarly stringent cost criterion – certainly not at such an early stage in their development' (Select Committee 1984: xxxii).

Of course, equally, it could be said that in the case of wave power the negative assessments were in the end right, although it is hard to prove that either way at present. In effect the jury is still out.

Given the uncertainties and regardless of the rights and wrongs of the way wave power was treated in the UK, it is obviously hard to decide how to develop new technologies. Much of the basic data is just not there and there are, especially with novel technologies like renewables, few precedents, and those that exist may be unhelpful. These problems are well illustrated by our second mini case study, concerning the way wind power was developed, focusing mainly on the approaches adopted in the USA, UK and Denmark. In this mini case study, our emphasis moves from the initial research phase on to the development phase and then on to commercial deployment.

The development of wind power

From the mid-1970s, the USA adopted a high-tech aerospace approach to wind power development, with the emphasis on increasingly large complex prototypes like the 2.5-MW Boeing/NASA 'Mod 2' series. Germany and Sweden also embarked on similar projects. When the UK started up its own wind programme in the early 1980s, it adopted a similar approach. After building a smaller 250-kW prototype, a large 3-MW machine was built on the Orkney islands off the north of Scotland by the Wind Energy Group. This eventually cost around £17 million.

However, the large machines were not successful: there were some technical failures due, for example, to the great stresses on the giant blades, and in general, after the initial enthusiasm, they were seen as too big, complex and expensive, pushing the technology too far too soon. Around the world commercial emphasis shifted to smaller machines of around 300–400 kW rated power.

Denmark had already taken a lead by initially emphasising relatively simple, robust, machines based on proven agricultural engineering approaches. Local agricultural engineers, operating on an almost craft basis, but with carefully targeted state support, developed small wind turbines, which subsequently proved to be world beaters. The Danish machines sold in great numbers to the USA and later to the UK and have been gradually scaled up as technical experience was gained and markets extended. Subsequently, Japan developed similar designs, as have the USA and the UK.

The argument against starting with large-scale projects in relation to wave power thus clearly applies also to wind power. Most of the large complex wind turbines initially developed in the USA (by NASA, Boeing, etc.) and in UK (by the Wind Energy Group consortium which included British Aerospace) and elsewhere (e.g. in Germany where a 6-MW unit was constructed) have now been abandoned.

Part of the problem with the top-down large-scale high-tech approach was that aerospace engineering concepts were not as relevant as expected when it came to devising systems that had to operate with dramatically varying wind loads. As one US wind turbine engineer put it, 'we were guilty of steady flow aerospace thinking, and largely did not appreciate the range and difficulty of the wind environment' (Stoddard 1986).

The eventual outcome illustrates the weakness of the research-led programme – adopting the so-called 'technology push' approach as opposed to being geared to responding to 'market pull'. Of course, in the early years there were no markets for wind turbines. But as the markets emerged, 'technology push' gave way to 'demand pull', and the Danes were fortunate to be in the right place to exploit this, for example on the basis of the rapidly expanding US market. For a while Denmark enjoyed a 90 per cent share of the US market. Subsequently, Japan entered the world market with devices of similar designs, followed by Germany.

Interestingly, machine sizes have increased dramatically with 1–2-MW machines now being in use, and some even larger units are being developed. Incremental development seems to be the order of the day. Certainly, the approach to initial machine development adopted in Denmark was far less expensive than that adopted in the USA – the Danish wind industry received a total of $52 million via government grants during the initial development phase, while the equivalent figure for the USA was $450 million (Karnoe 1990).

Overall it seems that the initial focus on small machines, developed on the basis of an incremental cut and try, 'bottom-up' approach, has clearly triumphed over the top-down high-tech approach and has laid the basis for expansion.

Technical success and failure

The wind turbine example shows how technological development patterns interact with market economic mechanisms and clearly economics is a central issue for renewable energy technologies. One way in which the institutional and economic problems can be reduced is if technological progress can be made, which may, for example, help reduce generation costs and build more confidence in the technology. Technological breakthroughs are possible, as are incremental improvements. Equally, however, there can be technological problems, which may dissuade funding sources from continuing to back a new technology. The risks may be perceived as too great.

This seems to have been the case in our next example – concerning the fate of geothermal 'hot dry rock' technology in the UK.

The UK geothermal programme

The potential geothermal resource is thought to be quite large, representing, if fully developed, perhaps 10 per cent of UK electricity requirements. The UK's hot dry rock programme involved an experimental well at Camborne, in Cornwall, funded by the government. A double well system was created, but the initial results were disappointing: less power was produced than expected due to problems with the geology. Establishing an efficient heat-extracting well configuration is a difficult and expensive art, even assuming the basic concepts are sound.

As you may recall, the basic idea is to create a fissure system between the bottom ends of the two wells to act as means of collecting heat, with cold water being pumped down one well and hot water/steam emerging up the other. The fissure system is usually created by exploding a small charge at the bottom of the wells. However, this has to be done in exactly the right way.

To put it simply, if the fissure system that is created allows water to pass through too easily, the flow through is too high and not enough heat is picked up. While if the artificial heat exchanger created by the fissure offers too much resistance to the water being pumped though, the flow rate is reduced and again not enough heat is absorbed. The fissure connections have to be just right and that depends on the precise geological nature of the strata. One might therefore expect failures before a successful well was achieved, much as oil exploration companies accept that many early drillings will prove fruitless.

Nevertheless, the initial failure of the Camborne well led to a loss of confidence in the project, which by this stage had cost some £42 million. In 1994, the UK programme was halted, with the government concluding that the technology was not likely to generate electricity economically. Work on geothermal power has continued elsewhere in the world, but the Camborne project has been portrayed as a failure (NAO 1994).

Conclusion

Technical problems like those experienced by the Camborne geothermal project are not uncommon at the early stage of the research and development process: initial technical problems are to be expected as part of the learning process. The withdrawal of financial support may have as much to do with short-term economic considerations as actual technological problems. It could be argued that a longer term perspective is required: the large geothermal resource around the world is unlikely to be ignored for long.

The same could be said in relation to ARBRE, the advanced wood gasification/combined cycle turbine project in the UK. As mentioned in Chapter 7, in 2002 this prototype plant met with some teething problems during commissioning, which led to a financial crisis when the developer decided to withdraw financial support. It may take some time for new technologies like this to become commercially viable, but it seems inevitable that, given the scale of the energy crop resource globally, new technologies will be developed to utilise it efficiently.

Fortunately, in most cases, renewable energy technology has been able to move up the learning curve, as illustrated by the case study on wind power. This has also happened in the case of photovoltaics and many other renewable energy technologies. Incremental bottom-up approaches, as adopted in Denmark for wind power, may not always be the best. For example, PV solar development has relied on high technology R & D. Moreover, sometimes there is no alternative other than for large top-down projects, as in the case of geothermal hot dry rock technology.

However, there are contexts in which incremental approaches may be best. While, as we have seen, the rapid development of large deep-sea wave power may have been stalled for economic and/or institutional reasons, smaller scale on-shore and in-shore wave energy devices are now being developed and seem likely to be successful. They could lead on to larger, fully offshore systems. It could be argued that the problems facing the ARBRE energy crop project could have been avoided if a more incremental approach had been taken, based on developing existing technology. This could also have provided an opportunity to develop a fuller relationship with local farmers, who were expected to supply wood chip for the plant.

Close involvement with 'users' is increasingly seen as important for successful technological innovation, since it can lead to incremental improvement in the design of the system. That is clearly what happened with wind technology in Denmark. Another example of this approach is provided by the successful development of solar power in Austria. In the mid-1980s, a grass roots solar heat collector initiative emerged. Local activists and enthusiasts, motivated by the need to get access to cheap and efficient solar heating technology, developed their own novel solar collector design. It was refined incrementally by the users within the enthusiasts' network, and was eventually taken up by 100,000 users, making Austria one of the leaders in the use of solar power.

Overall, one way or another, it seems likely that, in most cases, renewable energy technology will continue to develop in terms of performance and reliability, and become more cost effective. For example, the costs of electricity from wind turbines has fallen by around 70 per cent within a decade or so, and similar reductions for photovoltaic solar cells have taken place.

However, even if it is possible to obtain institutional support for research and then develop successful and economically attractive technologies, there are many other problems facing renewables in trying to become established as major energy options. The next chapter looks at the problems relating to the next stage in development – getting support for full-scale deployment. For some consumer-oriented products, like solar collectors, it may not be too difficult to get the idea taken up in the market place, as the Austrian example above illustrates, especially if a bottom-up approach is used. But, as we shall see, when it comes to larger and more complex technologies, the problems of getting support for full-scale deployment can be significant.

Summary points

- Obtaining support for research and innovation is often difficult.

- Vested interests may resist the development of new technological options.

- Incremental bottom-up development strategies may be more effective than top-down approaches.

- There may be lack of support for 'follow through' when projects are faced with initial problems.

- Even if new energy technologies can successfully pass through the research phase, there can still be problems with full-scale implementation.

Further reading

The UK wave power story is analysed in detail by David Ross in *Power from the Waves* (Oxford University Press, Oxford, 1995). The wind power story in the UK, USA, Denmark and elsewhere is covered by Paul Gipe in *Wind Energy*

Comes of Age (Wiley, Chichester, 1995). The UK Renewable Energy programme was reviewed by the independent National Audit Office in 1994. Its report includes useful critiques of the wind and geothermal programmes. Boru Douthwaite's *Enabling Innovation* (Zed Books, London, 2002) provides an excellent overview of what is needed to stimulate innovation, arguing that bottom-up approaches are usually far more successful than top-down approaches and using the Danish wind story as one example. For an account of the parallel bottom-up, self-construction solar power movement in Austria see http://www.aee.at/indexeng.html.

11 Keeping going: deployment problems

- The problems of 'short termism'
- The need for support infrastructure
- The need for public acceptance

Even when novel energy technologies have passed through the research phase successfully, it is sometimes hard for them to get support for full-scale implementation. The existing financial and institutional arrangements and priorities often do not favour new technologies. This chapter looks at some of the financial problems that have faced renewables, using tidal power as an example. It also looks at the problem of promoting energy conservation. Finally, it introduces the question of how public acceptance for novel energy technologies can be gained – since this is the ultimate hurdle in the implementation process.

The problems of transition

The momentum behind the development of sustainable energy technology seems fairly well established. For example, the economic benefits of energy conservation are clear and it seems likely that renewable energy technology will increasingly come to the fore as the environmental costs of existing energy technology become more apparent.

However, the pattern of development so far has revealed a number of major institutional problems that may make a rapid transition to a sustainable energy future difficult. These are not always trivial problems: they involve a battle over the whole paradigm of technological development.

In 1981, before his funding was cut, Stephen Salter, the wave energy pioneer, commented:

> We are attempting to change a status quo which is buttressed by prodigious investment of money and power and professional reputations. For 100 years it has been easy to burn and pollute. 100 years of tradition cannot be swept away without a struggle. The nearer renewable energy technology gets to success, the harder that struggle becomes.
>
> (Salter 1981b: 580)

Chapter 9 looked at some of the key elements of various possible strategies for sustainable energy development, and it was clear that there could be conflicts over the direction, balance and pace of development. Chapter 10 looked at some specific examples of these types of problems: it was hard for new technologies to get started, for example in terms of obtaining funding for research.

Unfortunately, even assuming that the necessary level of financial support for research and development is obtained, and that the technology can be developed successfully, there still remains a series of problems relating to the next stage – the practical full-scale deployment of renewable energy systems and energy efficiency programmes. These problems are the concern of this chapter.

Problems of deployment

A key problem facing new technologies and techniques is that they are inevitably trying to establish themselves in an institutional, market and industrial context based on the existing types of energy technology. There are powerful vested interests in the status quo, and in many cases, this is reflected in the current financial, organisational and institutional environment, which may not be very well suited to the acceptance of novel developments. There is, arguably, something of a mismatch between the new technology and the existing support infrastructure, and a disinclination therefore to provide the necessary funding.

To some extent this may be surprising. After all energy conservation has obvious economic advantages and, superficially at least, renewable energy is a free resource – there are, with most natural energy flows at least, no fuel costs. So they ought to be commercially attractive. However, all change involves disruption and the economic advantages may take longer to materialise than financial backers are willing to accept. As was illustrated in Chapter 4, the 'payback time' problem is often less in the case of energy conservation measures, but it can be serious for some renewable energy technologies. For although the fuel is essentially free, at least in the case of natural flow sources like wind or solar energy, there are, nevertheless, significant costs associated with constructing the necessary energy-conversion technology.

With conventional power plants, although the construction of the plant is expensive, a significant part of the overall cost is in the fuel, and this cost is incurred after the plant has been built. So some of the cost is deferred until after the plant has started earning its keep. By contrast, with renewables, since the fuel is free, at least with plants based on natural energy flow, there is only the upfront capital cost of constructing the system. So, rather perversely, it is harder to find investment sources.

Nowhere is this more clear than in the case of *tidal barrages*. Therefore, it is perhaps worth looking briefly at the fate of tidal power in the UK, which, as has been noted, has some of the world's largest tidal energy resources. Tidal barrages are not a new concept: the technology has been proved and a barrage has been operating successfully in France, on the Rance estuary, since 1968. However, obtaining finance for major tidal projects proved to be difficult in the UK.

Tidal barrages in the UK

Initially, when the UK renewable energy research and assessment programme started in the mid-1970s, the tidal option was seen as quite significant: the UK had some of the best sites in the world, notably the Severn estuary. A government-backed Severn Barrage Committee was set up to review the potential and it reported back in 1981, confirming the results of several earlier studies, which had indicated that a tidal barrage on the Severn estuary would be technically feasible and could generate power at competitive costs, once built.

Although the Severn barrage scheme was seen as technically viable, the capital cost of building it would be very large. The Severn Tidal Power Group, an industrial consortium consisting of many of the UK's leading construction and engineering companies, developed a proposal for an 11-mile long barrage, which would have 8,640 megawatts of installed generating capacity and would be able to meet on average about 6 per cent of UK electricity requirements. But the capital cost would be around £10 billion. Government-supported feasibility studies nevertheless went ahead, culminating in the Severn barrage project report in 1989 (Department of Energy 1989).

At this stage much of the emphasis was on the likely environmental impact of the Severn barrage scheme. Some environmental organisations had objected to it since they felt it could adversely affect wildlife and the local ecosystem generally, although, as the scheme's promoters pointed out, some of the social impacts could also be positive, for example, the provision of a new Severn road or rail crossing, increased opportunities for watersports and increased employment opportunities locally.

As it turned out, the main problem to face tidal power was not the potential local impacts on the natural environment but the changed economic environment that emerged following the privatisation of the UK electricity industry in 1989–90. The Central Electricity Generating Board, the UK's nationalised utility, was broken up and the smaller private companies that replaced it were unlikely to be interested in a major project of this sort.

A project on this scale might have been viable as a long-term publicly financed national investment, or perhaps via some partnership arrangement with industry, since public sector rates of return would then have been all that would be needed. However, following the privatisation of the electricity industry, it became clear that the government was not going to provide any further funding, and the Severn barrage project was stalled: the private sector would be unlikely to want to take it on single-handedly, since, in the economic climate of the time, it would be looking for much higher rates of return over shorter periods.

Attempts were made to investigate smaller, less expensive tidal barrage options, notably on the Mersey, and the Mersey Barrage Company, another consortium of major UK companies, tried to obtain NFFO support for their scheme. In the end, however, this attempt was unsuccessful, and the Mersey project and several other smaller barrage proposals, were abandoned. In 1994 the tidal programme was more or less wound up, after around £14 million had been spent (Department of Trade and Industry 1994b).

As can be seen, the primary problem in the case of tidal barrages was not technological: the technology existed and was relatively mature. The problem facing the UK barrages was financial, and the relatively short-term economic perspective that prevailed in the energy sector, as elsewhere, following the privatisation of the UK electricity sector. Once built, barrages would pay off their investment costs in a matter of decades, after which, like hydroelectric plants, they would produce very cheap electricity for centuries. However, the initial capital cost is large, and since the technology was relatively mature, significant cost reductions were unlikely. As a result, investment funds proved impossible to find: the payback times were simply too long.

Supporting innovation

Large projects like tidal barrages clearly have problems in obtaining funding. But although smaller scale renewable energy products may have the attraction that they can be deployed incrementally on a modular basis, similar problems can also face them. The problem is essentially one of institutional preconceptions, for example, as to what represent suitable areas of technological investment by financial institutions. In the past, the emphasis in terms of major financial investment has been on the use of large-scale concentrated forms of energy, managed by large-scale centralised agencies. Investment agencies are therefore often suspicious of smaller projects, which are sometimes viewed as likely to be risky, low-yield investments.

This has been particularly true in the UK, where some wind projects have found it difficult to get finance, given the high rates of return expected by private sector investors, and the risks they feel that these novel technologies involve. Certainly, some of the pioneers in the renewable energy field have found it hard going. For example, the first wind farm in the UK was established in 1991 by Peter Edwards, a Cornish farmer and landowner. The project was given support under the government's non-fossil fuel subsidy scheme, but this would only provide an income when power was produced: it did not provide 'up-front' money for construction. He had the benefit of a 40 per cent EU capital grant, but this was conditional on finding the rest of the capital elsewhere, which he achieved in part by selling off his dairy herd.

The UK government's introduction, as described in Chapter 8, of the Climate Change Levy in 2001 and the Renewables Obligation in 2002, should in principle have improved the situation for UK renewable energy project developers. However, in practice the positive impact of these market drivers has been undermined by another government initiative, the *New Electricity Trading Arrangements* (NETA), which was introduced in

2001. These aimed to increase competitive pressures amongst energy suppliers, especially the large suppliers, in order to force prices down. It has done that very successfully. Wholesale electricity prices fell by 20–25 per cent in the first year of NETA. However, the impact on renewable energy schemes was disastrous – most are small companies that could not compete with the large conventional generation companies when prices were pushed down. The result was that demand for their output fell by 44 per cent.

Part of the problem was that, as we noted in Chapter 9, individual renewable energy projects like wind farms can only offer power intermittently and this is penalised heavily in the NETA structure, despite the fact that the grid control engineers cannot see the variations in the outputs of the wind projects, which are lost amongst the much larger variations from conventional sources. Moreover, the collective environmental contributions of the wind projects, in terms of reduced carbon emissions, are not recognised. Neither is the fact that they mostly deliver power locally, thus avoiding power losses on long-distance transmission. The end result was that some wind farm projects had to close down and the government was pressed by the various energy interests to modify the NETA system.

The pressure for changes in NETA was not just due to the problems faced by renewables. CHP plants were also facing difficulty. Moreover, as mentioned in Chapter 6, British Energy, the UK's main nuclear plant operator, also fell foul of the new, very competitive climate and had to be bailed out by the government. Some coal projects also faced problems. With major suppliers finding it hard to keep going, the situation could develop to have some parallels with that in California, although so far, in the UK, consumer prices have not been affected and supplies have not been disrupted, in part because the UK has around a 30 per cent excess of generating capacity over peak demand requirements. However, the situation was clearly worrying, not least for renewable generators. If well-established technologies like nuclear and coal were having problems, then new entrants like renewables would find it difficult to survive.

In some other countries the institutional and financial support infrastructure are better matched to the needs of new projects, and this may provide an indication of how progress can be made in future. For example, as has been illustrated earlier, wind power is very successful when it is developed and deployed on a small, local scale, as in Denmark. As you will recall, the locally produced Danish wind turbines turned out to be world beaters, selling in large numbers around the world. Perhaps equally important is the fact that this export success was launched from a strong domestic market, which had been created from the bottom up. As noted earlier, around 80 per cent of the wind turbines installed in Denmark so far are locally owned by individuals or co-operative guilds. Local ownership and local development has meant that funding was relatively easy to obtain: local banks were willing to provide loans.

Interestingly, another major renewable energy source, fast-growing energy crops like willow or poplar trees, might also be best developed at the relatively small-scale local level. The ARBRE wood chip gasification project mentioned earlier may perhaps have been too ambitious technologically for the first major energy crop utilisation project in the UK, but it did pioneer the establishment of a support network for energy crop supply, involving a farmers' co-op. Smaller scale projects, using more conventional combustion technology, could develop this idea and provide the context for the incremental development of more advanced technology. Certainly, energy crop growing seems likely to fit in well with existing rural agricultural economic and social patterns and practices. For example, banks are accustomed to providing loans to farmers and the technical infrastructure for growing and harvesting crops already exists on farms.

This discussion highlights a general point that has emerged from current theorising on innovation: the successful deployment of new technology requires the existence or the development of suitable social and institutional contexts – a technical infrastructure, suitable financial networks, a skill base and local support. That may not yet exist for all of the renewables, but for some of the smaller scale, modular and locally scaled options, it seems to be emerging.

Supporting energy conservation

The same points also apply in the case of energy conservation, possibly with even more force. As has been noted, there can be problems with deploying energy efficiency measures, for example, due to the fact that they often require responses by many independent consumers. In this situation, it is vital to have support networks, providing advice and, where necessary, offering grant aid to stimulate uptake.

There have been many attempts to establish such networks and support schemes. There is also a range of institutional mechanisms, such as integrated resource planning, which can be used to try to ensure that, when investment decisions are being made by companies, the relative merits of investing in energy saving and investing in energy supply are given equal consideration. This approach has sometimes proved particularly fruitful in countries like the USA, where some power companies have found that they can obtain better financial returns by selling energy efficiency packages to their consumers than by investing in new energy supply technology (Beggs 2002).

However, in many countries, energy conservation initiatives still tend to be given low priority. This has been particularly so in the UK, where, following the privatisation of the energy supply industry, the institutional structure of the supply side means that it was in the interests of the various generating and distribution companies to sell more of their products, in competition with each other. Attempts have been made to try to resolve this problem by introducing 'standards of performance' in terms of the energy efficiency of consumer electrical goods, with some success. Improved building regulations have also begun to help and certainly there is a lot that can be done by government regulatory efforts of this sort. In addition, there is a need for sensible planning guidelines, to try to ensure that the design of the built environment and the associated pattern of commercial, industrial, residential and recreational development reflect concerns about energy use, for example, in relation to transport.

More directly, governments can also change patterns of energy use, and stimulate the uptake of energy conservation measures, via *energy taxation*. Many environmentalists feel that energy is basically too cheap at present – the price does not reflect the full environmental costs. However, the prospects for a serious approach to energy taxation are poor, given that such taxes are politically unpopular. For one thing energy taxes are economically regressive – they hit the poor hardest. However, most of the opposition to energy taxes has been less altruistic. The energy tax proposed at one time by the European Union was abandoned after political pressure from, amongst others, the UK government, which was concerned about the reaction of consumers. Certainly, the UK government's attempt in 1995 to impose the full level of value added tax (17.5 per cent) to domestic fuel resulted in a massive public backlash. In 2001 the belief that vehicle fuel taxes were too high led to blockading of petrol depots by protestors and a major UK political crisis.

Energy taxes that have specific environmental goals, rather than just being seen as general taxation, may be more popular. For example, the UK has imposed a form of carbon tax on vehicles, linked to engine size and, as noted in Chapter 8, a Climate Change Levy has been imposed on the power used by businesses, with power from renewable sources being exempted, in both cases without too many negative reactions. Energy supply companies will pass the cost of meeting the Renewables Obligation on to domestic consumers (rising to perhaps 4 per cent extra by 2010), which may act as a slight stimulus for energy conservation. But, otherwise, for the moment, in terms of energy used by the domestic consumer, the emphasis in the UK still remains basically on voluntary efforts, backed up in some cases by grants for specific projects, and stimulated and supported where possible by advice agencies.

It is interesting in this context that one result of the battle that started in 1995 over VAT on fuel in the UK was that the VAT charge was eventually removed from energy conservation materials and devices, including, subsequently, some domestic-level renewable energy systems, in the expectation that this would stimulate uptake. A network of energy advice centres was set up around the country to aid this process, and throughout the UK there are many voluntary organisations providing information on energy saving at the community level.

Social acceptance

The existence of advice centres, together with technical and financial support networks and schemes, is not something to be taken for granted. They are vital as a practical way to support the uptake and deployment of sustainable energy technologies, and also as a way to try to ensure that new often unfamiliar technologies are accepted by the general public.

Social acceptance is particularly important in the case of energy conservation initiatives in the domestic sector, since their success depends on the adoption of new technologies and techniques by consumers. Information and advice centres can alert people to what is available, but uptake has often proved to be a problem: there is still evidently a need to convince consumers of the benefits and then to help them to choose the best options. A National Opinion Poll carried out for the Energy Saving Trust in 2001 found that, of those asked, 73 per cent said they wanted more information about *why* they should save energy, while 77 per cent said they wanted information on *how* to save energy.

At first sight, social acceptance would seem less of a problem in the case of renewable energy technology. Opinion surveys over the years have indicated widespread and continuing popular support for the development of renewable energy technology in general terms. For example, 87 per cent of the respondents to a Gallup poll in the UK, carried out in 1991 for Friends of the Earth, indicated that they would prefer the government to increase spending on renewables. Ten years later the National Opinion Poll for the Energy Savings Trust, which was mentioned above, found that, of those asked, 85 per cent wanted the government investment in renewable energy. Even so, when it comes to specific projects there may be objections, with adverse local environmental impacts being a key issue.

In principal, compared to most conventional energy generation technology, renewable energy systems like wind turbines ought to have some advantages in terms of public acceptability. They are relatively small scale and although there may be a need for fairly large numbers of them, in contrast to the hidden dangers of, say, a nuclear plant, they have the advantage that 'what you see is what you get' – they are functionally transparent. Their purpose and operation is usually clear from their appearance and there are no hidden or longer term environmental problems. For example, if necessary, wind farms can be easily decommissioned and removed, thus returning the site to its original state.

However, the widespread introduction of renewable energy technology like wind turbines could involve significant changes in land use and landscapes and might even involve changes in social patterns, for example, a degree of decentralisation. So if renewables are to develop on a significant scale then it is vital that they gain public support.

Put less passively, it seems vital that the public can influence the way they are developed and deployed. The early enthusiasts for 'alternative technology' argued that one of its attractions was that it could be developed and used on the smaller scale, and might therefore be more susceptible to democratic control on a local level. Not all renewable energy systems fit this description, but some do, and it will be interesting to see what role local control can play in reality in shaping the way renewable energy develops.

This process may not always be easy or without conflict, for example, in terms of local environmental impacts. These conflicts provide an interesting example of the problems of 'thinking globally and acting locally'. Certainly, the development of some renewable energy technologies is likely to be constrained by local environmental, planning and land use factors.

Equally, the deployment of renewables may be stimulated by increasing environmental concerns over the generally much more significant global impacts of using conventional energy technologies, for example, global warming from the emission of greenhouse gases like carbon dioxide produced when fossil fuels are burnt. The local and global impacts have to be traded off against each other.

Environmental conflicts

Economic concerns inevitably determine the success or otherwise of technological projects, but land use issues and environmental concerns also enter into the commercial equation – if for no other reason than because it takes time and money to obtain planning permission. So while conventional economic factors are obviously important, the need for a trade-off between local and global environmental factors is also important and may shape the way in which renewables are developed.

Renewable energy provides part of a solution to global environmental problems, but no technology can be entirely benign in environmental terms. Although the impacts of most renewables are relatively small and localised, compared to the often large and global impacts of conventional energy technologies, there can still be local problems. Perhaps the most significant impacts are associated with large tidal barrages and hydro plants, which involve large civil-engineering constructions and large modifications to local and even regional or national ecosystems and natural energy flows.

Studies of public attitudes to proposals for a barrage on the Severn estuary in the UK have indicated some mixed responses (Barac *et al.* 1983). There was a general enthusiasm for renewable energy projects, and support for the local economic benefits that the advent of large projects like this could bring, not least local employment. Equally, however, there was concern over local impacts on wildlife and the ecosystem generally, combined with more instrumental concerns about any adverse effects on the ease of navigation by shipping.

Although wind farms involve less environmental modification, they also intrude on the landscape. The first wind farms, installed in California in the late 1970s and early 1980s, were generally well received, but there were also some objections. Many of these related not so much to the visual appearance as to concerns that the machines were not really economically viable. There were suspicions that they were being installed without much concern for their technical success, simply to take advantage of the tax credit system being operated by the Californian state government as an incentive for wind energy development. Certainly, many of the early wind projects turned out to be technically unviable, and the planning controls on their siting were, in some situations, fairly lax. In essence, what had happened was a 'wind rush', somewhat like the 'gold rush' a century or so earlier. This rather chaotic phase nevertheless had some benefits: it provided a context in which new designs could be tested and certainly the technology developed rapidly. So too did awareness of the need to win public acceptance. In some cases this required no more than the provision of better information. But it also became clear that there was a need for careful siting and layout to minimise visual disruption and for sensitive local consultation (Thayer and Hanson 1988; Gipe 1995; Pasqualetti *et al.* 2000).

As the case study in the next chapter illustrates, this became even more clear when the first wind farms were built in the UK, where population densities are generally much higher. The problem of noise pollution also became important.

Technological developments are helping to reduce some of the problems. For example, the new generation of variable speed wind turbines are less noisy. But it will still be important for developers and planners to be sensitive to local concerns and to consult with local communities over the location and layout of proposed schemes.

To summarise, renewable energy technologies present system designers and planners with an interesting challenge: they must try to balance the global advantages of renewables (in terms of, for example, reduced emissions of greenhouse gases like carbon dioxide) against the local impacts, and come up with technically workable, economically viable and environmentally acceptable compromises. What seems to be needed is some way to negotiate a balance between global and local.

The wind farm case study

The next chapter present a detailed case study on public reactions to wind power, which in effect represents the 'first shots' in this process. The aim of the case study is to provide an insight into how the political process of conflict and negotiation discussed in Chapter 1 works, or might work, in practice.

Case studies could have been chosen from many other areas, for example based on local concerns around the world over emissions from waste combustion plants, the conflicts that have emerged in some locations over the development of geothermal energy systems, or the debates on ecosystem impacts in relation to large-scale hydroelectric projects, or tidal barrages. But the debate over wind farms has the merits of being well documented, particularly in the UK, and it involves a smaller scale technology whose nature and likely impact is relatively easy to understand.

Summary points

- Some large renewable energy projects face an uphill battle in terms of obtaining finance for full-scale deployment since the benefits are sometimes longer term.

- Smaller scale projects may be able to obtain local finance and be more publicly acceptable.

- There is a need for public awareness of the relative costs and benefits of sustainable energy technologies.

- Public concerns have to be addressed and local reactions have to be considered in the process of developing sustainable energy technologies.

Further reading

There has been a long-running debate over the impact of renewable energy, particularly in relation to wind farms and tidal power. The pros and cons of tidal power are discussed by Clive Baker in *Tidal Power* (Peter Peregrinus/IEE, London, 1991). See also Carol Barac, Liz Spencer and Dave Elliott, 'Public awareness of renewable energy: a pilot study', *International Journal of Ambient Energy*, 4(4): 199–211, 1983, which looks at reactions to the proposed Severn barrage.

The UK wind farm debate is the subject of Chapter 12. A useful starting point for a discussion of reactions to wind farms in the USA and elsewhere is provided by the book edited by M. Pasqualetti, P. Gipe and R. Righter, *Wind Power in View: Energy Landscapes in a Crowded World* (Academic Press/Elsevier, London,

2000). The classic UK study of impacts is Alexi Clarke's report 'Wind farm location and environmental impact' (NATTA, Milton Keynes, 1988).

The Energy Saving Trust is a useful source of information on the problems facing energy conservation in the UK, and what is being done about it: see http://www.est.co.uk.

Case study: public reactions to wind farms in the UK

- Conflicts between developers and local communities
- The role of planning
- The need for sensitive consultation and siting

New energy technologies are inevitably unfamiliar and their deployment can lead to public concern – particularly if they are perceived to have negative local impacts. This chapter presents a case study on the often heated debate over the location of wind farms in the UK. It provides an example of the need, when seeking to introduce new projects, to be sensitive to local concerns and at the same time to try to balance local environmental costs against global environmental benefits.

Introduction

The UK wind farm programme, which got underway from 1990 onwards, owes its existence primarily to the non-fossil fuel cross-subsidy scheme introduced following the privatisation of the electricity supply industry.

As was noted in Chapter 8, a Non-Fossil Fuel Obligation (NFFO) was imposed on the newly privatised regional electricity supply companies (RECs), requiring them to buy in set amounts of electricity from nuclear and, to a much lesser extent, renewable suppliers. In addition, a surcharge was imposed on fossil fuel electricity generation in order to meet the extra cost of using non-fossil sources, with this cost being passed on by the RECs to their customers.

Two NFFO orders were set specifically for renewables in 1990 and 1991, with in all 197 projects being offered a 'premium' price, over and above the usual 'pool' price, for their electricity. The structure and constraints of these NFFO orders played a major role in shaping the initial pattern of wind farm development and, arguably, public reactions to it. The central problem was that the NFFO cross-subsidy scheme was seen as likely to be an infringement of the EC's rules on fair competition. Some EC members were also opposed to providing support for nuclear power. Consequently, as a compromise, a 1998 deadline was imposed for the NFFO scheme, with renewables, in effect, being inadvertently penalised. As we shall see, subsequently, for NFFO 3 and thereafter this constraint was removed for renewables. However, the 1998 deadline played a significant role in shaping the initial development of wind power in the UK: it meant that intending developers could only receive the premium cross-subsidy price for a limited period, which was reduced as the cut-off date approached (Elliott 1992).

The result was something of a 'wind rush', with perhaps insufficient time for full environmental assessment. The 1998 cut-off date also meant that the premium price offered had to be quite generous and artificially enhanced. Wind projects were offered 6p per kWh in the 1990 NFFO and 11p in the 1991 NFFO, in order to allow companies to recoup their investment in the shorter period remaining. This had the effect of making wind power look very expensive and raised the spectre of 'profiteering'. Wind farm developers were sometimes seen as rushing in to exploit 'handouts' without paying sufficient attention to environmental impacts.

There may be some truth in this: certainly some developers with good windy sites may have benefited from the fact that all the projects received the same basic average payment regardless of the site. Equally, however, it could be argued that developers had little choice but to try to find windy, upland sites. Even given the relatively high NFFO levy payments, for most developers the 1998 cut-off date meant that it was hard to get financial backing. Their profit margins were tight and most evidently felt compelled to target the high wind speed sites. This would have been financially tempting in any case, but the 1998 deadline provided a further impetus. As we shall see, given that these upland sites were sometimes in environmentally sensitive areas, this led to significant local objections.

Reactions to wind farms

As the practical deployment of wind projects picked up speed from 1991–2 onwards, data on public reactions became available. Before and after responses to the first project – a 10 Vesta turbine scheme at Delabole in Cornwall – were carried out by consultants for the government's Energy Technology Support Unit (ETSU). The 'before' study was carried out in 1990, the 'after' in mid-1992, six months after start up. A 'control' study of opinion in Exeter was also carried out.

The conclusion was that basically wind power was popular. Only about a third offered NIMBY (not in my back yard) responses and support for wind farms consolidated when people had experience of it locally. For example, around 25 per cent of those who were concerned initially about the Delabole scheme changed their minds: 80 per cent said it had made no difference to their day-to-day life; 44 per cent approved; and 40 per cent approved strongly. On visual intrusion, more than 40 per cent had thought it might be a problem in 1990, whereas this had fallen to 29 per cent by 1992. On noise, only 14 per cent had not expected this to be a problem in 1990, but by 1992 around 80 per cent felt it was actually not a problem.

Overall, the introduction of the wind farm had 'altered attitudes in the direction of local residents being more favourable towards wind energy' with 'many of the worries local residents had about wind turbines' having been proved unfounded (Energy Technology Support Unit 1993b). Interestingly, Delabole had around 100,000 visitors in its first year.

Subsequent studies have shown similar patterns of support. For example, wind farm developer National Wind Power reported strong local support for its wind farm at Cemmaes in mid-Wales: 98 per cent of local people asked 'liked or didn't mind' it, while, according to Ecogen, another development company, a survey of 500 local adults, carried out by a local school near the Coal Clough wind farm, found that 70 per cent wanted more wind projects. Interestingly, it also found that, although 40 per cent of the people sampled could not see the wind farm from their homes, 55 per cent of them would have actually preferred to be able to see it. Positive reactions also emerged from a survey of 1,000 residents by a local Friends of the Earth group in Sidlesham on the Manhood Peninsula: 83 per cent supported the wind farm proposal there; 7 per cent were against; and 10 per cent were unsure.

However, wind projects were clearly not popular with everyone. There were some bitter planning battles as the UK programme got underway, with, as we shall see, local opposition to some projects becoming very significant.

The planners' and conservation groups' responses

The first wave of wind farm projects faced local planners with considerable problems. There were few precedents for this type of development, for example, were they to be seen as agricultural projects as reflected in the label 'wind farm', or were they industrial projects, that is, power plants? Consequently there were requests for new planning directives from the government. Draft (consultative) Planning Guidelines emerged from the Department of Environment in 1991. However, in the main, they left the detailed assessment up to the local planning authority; the basic policy being to support wind for global strategic reasons unless local costs outweighed them. In effect, they asked the planners to balance the national and global benefits against any local disbenefits.

However, local council planners were in general unhappy with the lack, as they saw it, of clear guidance. Thus, responding to the Draft Planning Guidelines, the Association of District Councils suggested that without 'clear statements of strategic need, the Local Planning Authorities may lack a sufficiently strong case to justify the inclusion of positive proposals for wind farms . . . in local plans, against the probable weight of local objections' pointing out that 'Local planning authorities are not responsible for energy generation' (Association of District Councils 1992).

The County Planning Officers Society added 'it is doubtful if individual local planning authorities can realistically be expected to consider the overall environmental benefits when the actual rewards remain intangible' (County Planning Officers Society 1992).

Some conservation and environmental groups went further and some, like the Countryside Commission and the Council for the Protection of Rural England, felt that the Planning Guidelines indicated that the government was being 'soft' on developers. By contrast, some of the more radical pressure groups saw the problem as being not so much lack of control of and/or profiteering by the developers but a failure by the government properly to support wind power. Thus the Campaign for the Protection of Rural Wales indicated that, while they would oppose any specific project they felt was ill-conceived, they would also 'lobby hard to amend the financial package currently on offer to developers', so as to allow for less invasive siting (Evans 1991).

For its part, the government tried to avoid making a link between planning and financial issues. Thus, in a letter to Friends of the Earth (28 March

1991), Colin Moynihan, then an energy minister, denied any link, claiming that the siting issue was 'fundamentally a planning issue rather than a commercial one'.

The full Planning Policy Guidelines (PPG22) emerged in February 1993, but they differed only in detail from the draft, and did not really resolve this issue. Responding to continued pressure for clarification, in November 1993, Tim Eggar, an energy minister, spelt out the government's position as follows:

> The Government does not have a specific target for wind energy, and its success will depend on developers finding sites acceptable to the public and to planning authorities. NFFO does not override the planning process and Government is as concerned as much about the local environment as the global one. The planning guidance in PPG22 requires planners to balance the Government's policies for renewables with those of the countryside.
>
> (Eggar 1993)

Making this type of judgement is obviously difficult and, in principle, each case was to be judged on merit, without setting a precedent. Nevertheless, the first few planning inquiries did seem to follow a common pattern.

Public inquiries

The first Public Inquiry, in 1991, was on the Wind Energy Groups 24-turbine wind farm proposed for Cemmaes in the Dyfi Valley in mid-Wales, on the edge of Snowdonia. Despite objections by the Countryside Commission, amongst others, it led to very positive recommendations from the inspector, whose conclusion was subsequently accepted by the then Secretary of State for Wales, David Hunt. Although the visual impact issue was seen as relevant, Hunt felt that it was 'not sufficiently compelling to outweigh the need for renewable energy' (Hunt 1991).

The next inquiry focused on WEG's proposed 15-turbine wind farm on Kirkby Moor in Cumbria, just south of the Lake District National Park. The inspector turned it down on the grounds of visual intrusion, but his conclusions were overruled by the Secretary of State for the Environment who argued, 'such harm as may be caused by the visual impact of the wind farm in this instance is outweighed by the national need for the development of alternative cleaner sources of energy' (Department of the Environment 1992).

Most of the applications, of course, were not called in for full planning inquiries, and, at least initially, the majority of them obtained planning permission, even if, in some cases, this had required the intervention of the Secretary of State. For example, by the end of 1993 John Gummer, the Environment Secretary, had overturned four rejections by local planners. However, this was not a fixed pattern. After strong local opposition, the Welsh Office turned down an appeal relating to an application for a National Wind Power wind farm on Anglesey; and more than fifteen other proposals have been turned down by local planners. Even so, the objectors, and those environmental groups who were opposed to the wind farms, evidently felt that they were fighting an uphill battle.

Thus a local solicitor, who had acted for objectors to Ecogen's proposed wind farm on St Breock Down in Cornwall commented 'everyone believed it would be unthinkable in the face of opposition from just about every quarter. The inspector just steam rollered over the objectors' (Key 1994).

The environmental groups' response

The response from the environmental groups was mixed. Friends of the Earth, locally and nationally, maintained their long-held support for and promotion of wind power. So did the Labour Party-affiliated Socialist Environment and Resources Association (SERA) and Greenpeace. The WWF and the Royal Society for the Protection of Birds (RSPB) were also supportive. However, as we shall see, some other major conservation groups, such as the National Trust, were more critical of wind farms, while the Campaign for the Protection of Rural Wales, which initially adopted a supportive if critical stance, subsequently changed sides. So did the Ramblers Association, while the Welsh Tourist Board's 1994 'Tourism 2000' report expressed concern over the impact on tourism.

The Countryside Council for Wales (CCW) came out with a particularly strong opposition line arguing that while wind turbines 'are welcome as a source of renewable energy, the scale of their contribution to meeting energy needs does not justify overturning established planning policies and safeguards' with wind power projects tending 'to threaten precisely those areas that CCW is charged to protect'. They also added that there 'should be a presumption against wind turbine development in areas of close proximity to sites with the benefit of statutory landscape designation status' (CCW 1992).

Their English equivalent, the Council for the Protection of Rural England (CPRE), also expressed critical views. For example, in its 1991 evidence to the Department of Energy's Renewable Energy Advisory Group, CPRE called for greater scrutiny of projects and suggested that wind power should not be seen as a technical fix for 'the key political, social and economic problem – the profligate use of energy' (CPRE 1991).

Subsequently, Tony Burton from the CPRE told the *Guardian* (11 March 1994) that while they were not opposed to wind power in principal, 'the system of subsidies is putting pressure to build wind farms in quite inappropriate places, remote landscapes that have been protected for decades'.

To summarise then, while some environmental groups supported wind power; some moved to oppose it, and some evidently felt that, in the absence of proper government policies, they were having to 'police' the developers' projects unaided.

For their part some local planners resented being asked in effect to make decisions that were related to national energy policy, as they saw it, without proper guidance. In response, some have argued that it is beyond their competence and remit, and they therefore focus only on local issues. This has resulted in the rejection of some wind farm proposals, which subsequently may have been overturned.

That is not to say that local planners or all the conservation groups were necessarily against wind farms in principle. The objections were to the pace of development and lack of guidance.

The mood of some planners was well expressed by Tim Horwood, planning officer for Cornwall County Council, which produced its own interim policy guidelines: 'Instead of so many schemes opening at once it would have been better if they could have evolved more slowly so there would have been more time to consider all the implications' (Horwood 1992).

The wind backlash

Given that relatively large numbers of project proposals were coming forward, it is perhaps not surprising that some conservation groups moved into opposition. Some were clearly concerned that the programme was moving too fast. Thus the Northern Devon Group of the CPRE saw it as a 'mad scramble' for lucrative hilltop sites, with wind turbines threatening 'to stride across the countryside of North and West Devon like a plague of triffids' (Allen 1991). Barry Long from the Countryside Council for Wales told *The Times* (21 August 1991), 'It will be hard to stand on a hilltop in mid Wales without seeing a windmill'.

During 1993 objections emerged around the country, for example, in Devon, Cornwall and Yorkshire. But perhaps the largest number emerged in Wales. The UK's largest wind farm, so far, at Landinam, proved to be something of a turning-point in the debate. There had been objections to its scale, on visual intrusion grounds, but in the event, it was noise that proved to be the major problem. Several local residents claimed to be suffering from major disturbance, and there does indeed seem to be a significant noise problem for some residents in the valley below the ridge on which the 103 Mitsubishi machines are sited (Walker 1993).

The local experience with this project stimulated significant objections to subsequent projects in Wales, for example, in relation to National Wind Power's (eventually successful) application for permission to install a 22-turbine wind farm at Bryntitli. And it helped consolidate local opposition generally. A Noise Action Group was formed locally, and subsequently a National Windpower Consultative Association has been set up, initially covering the Welsh border areas.

An extensive debate ensued in the local press, with opinions often becoming polarised. The local responses have occasionally thrown up some bitter invective. One unsigned leaflet, circulated in mid-Wales in 1993, talked of Wales being 'in the forefront of being covered in swathes of ugly turbines to line the pockets of foreigners and greedy owners' (NATTA 1994a), although, in general, the debate has been carried out in less aggressive terms.

Even so, some colourful allusions have emerged. The Ecogen wind farm at Llandinam was claimed by a resident to sound like a 'twin-tub' washing machine. National Wind Power's wind farm at Llangwyrfon was alleged to sound like 'an old wheelbarrow being pushed along continuously', while their Cold Northcott project in Cornwall was described as sounding like 'a huge washing machine gathering speed to spin dry' (NATTA 1993a).

Clearly, noise was the major issue for these people, and this is a difficult matter to address. Visitors are usually surprised at how quiet wind farms sound, just a slight blade swish even close up, together with occasional gear train rumble. However, some machines are noisier than others and in some topographical situations these sounds can evidently be amplified by resonance effects, for example, by valleys. People's responses to the result can also vary. Some are very sensitive to low-grade background noise (and cannot sleep with a fridge running). Certainly, once a noise starts to be annoying, it can be detected even at very low levels. Some individual machines may have been particularly noisy during their run-in periods. The developers have been trying to respond to noise problems generally, for example, by installing noise insulation materials in the housing of the machines. More recently the development of more advanced variable speed direct-drive machines has significantly reduced noise levels, since there are no longer any gear trains to rumble, and movement of the blades is better matched to the airspeed, so there is less noise from air turbulence.

Less can be done by the designers and developers about visual intrusion, and this has become the main problem. 'Lavatory brushes in the sky' was how Sir Bernard Ingham described the wind farm near Hebden Bridge in Yorkshire, an allusion which has subsequently been repeated in various forms by the media. As one-time press secretary to Margaret (now Lady) Thatcher and public relations adviser to British Nuclear Fuels, Ingham has extensive media contacts and has been very outspoken on the wind farm issue.

With the involvement of such major public figures, the wind farm issue began to take on a national perspective and gained considerable national media coverage. Perhaps the key event was the setting up early in 1993 of a national anti-wind lobby group, Country Guardian, dedicated to opposing 'the desecration of the coasts and hills by wind farms', with Sir Bernard Ingham as vice president. At the start of the campaign, Joseph Lythgoe, who set up Country Guardian, sent out 1,800 letters, 'one to every weekly newspaper in the country'. Since then the Country Guardian group has been very successful at gaining media attention.

Subsequently, a campaign against the proposal to site forty-four turbines at Flaight Hill, near Hebden Bridge, has also attracted national media attention and the involvement of a number of celebrities, including pop star Cliff Richard and many notable literary figures. They wrote a letter, with sixty-two signatories, to the *Times Literary Supplement*, complaining about what they saw as an 'assault on our literary and artistic heritage', given that the wind farm would be in Brontë country.

The media's response

The press generally showed considerable interest in the local debate over wind farms. All of the major national newspapers (*The Times*, *Guardian*, *Independent*, *Telegraph*, *Observer*, *Financial Times*) have carried reports, with coverage increasing as objections mounted and the emphasis mainly on the negative side.

The broadcast media also have generally adopted a fairly critical and, in some cases, hostile approach. The BBC's 'Country File' programme (24 April 1993) included a quite critical review, while BBC Radio 4's 'You and Yours' programme (30 July 1993) presented a more or less unremittingly negative view. A Radio 4 'File on Four' programme (8 March 1994) was a little more hopeful, although it did suggest that the development of wind farms might turn out to be harder than the developers, and some environmentalists, had initially hoped.

The local press in the relevant areas carried regular news stories and features plus extensive letters. A rough survey of local press coverage in mid-Wales during the period between March and December 1993, although in no way exhaustive, may be indicative of the general pattern: there were sixty-two news items, nearly all reporting 'problems', and thirty-eight letters, with only eight being pro-wind (NATTA 1994b).

Most of the objectors cited specific local problems – noise and visual intrusion, but some reflected wider conservation and preservation concerns, as well as fears about the impact on tourism. Some of the supporters complained about the 'lack of balance' in the media debate, with Ian Mays from the British Wind Energy Association complaining to the *Guardian* (5 November 1993) that 'a small but vociferous number of people have generated a disproportionate amount of press coverage'. Nevertheless, the campaigns clearly had some effect. According to the *Guardian* (9 March 1994),

> Tim Eggar, the Energy Minister, is believed to be alarmed at the number of objections to proposed wind farms by groups who claim that their turbines impose noise and visual blight on the landscape. The DTI has been inundated with protests as part of a campaign at local and national level orchestrated primarily by . . . the Country Guardians.

A DTI spokesman had earlier denied that any fundamental shift in policy was likely, telling the *Financial Times* (22 December 1993) that, 'we don't need to radically change our policy because of opposition, because we have always said that the go-ahead for developments is dependent on the relevant planning permissions'.

Even so, it seemed likely that wind projects were going to be much more carefully scrutinised in the next round of the NFFO. NFFO 3 had been heavily oversubscribed, with over 650 renewable energy project proposals, including many wind projects. On 11 March 1994, just after the deadline for final applications had passed, Tim Eggar commented:

> Although some 230 of these proposals are for wind energy I would expect to see no more than twenty or so wind farms result from the next round of the NFFO. But this will depend on developers' abilities to find sufficient windy sites which are acceptable in planning terms, particularly from the point of view of noise and visual impact. If they fail to do so, or to get nominated projects up and running, this will count against wind energy in future rounds.

Reflecting the media campaign, he added,

> 'I can understand the concern expressed by some about the number (of) wind farms submitted for the latest round of contracts. I am prepared to see the steady development of wind energy, but it is not the case, as suggested by some alarmists, that I intend to cover the country with wind farms. The holding of a NFFO contract does not bestow any special consideration. Government planning policy requires the environmental benefits of wind and other renewables to be taken into account, but also places just as much emphasis on the need to protect the local environment.
>
> (Eggar 1994)

The validity of the objections

Mike Harper, then director of the British Wind Energy Association, asserted that 'the controversy to date has largely revolved around misconceptions and misinformation distributed by groups aiming to stifle wind energy development completely' (Harper 1994).

Certainly, there have been cases of misrepresentation and even disinformation; for example, the 10 per cent fossil fuel levy has sometimes been cited by wind farm objectors as being the extra cost imposed on electricity consumers by the wind farms. In fact the bulk of this 10 per cent is due to the support provided for nuclear power. Even given the initial artificially high level of support that had to be provided to wind projects as a consequence of the 1998 NFFO deadline, the wind farm element so far has still only added around one tenth of a per cent extra on an average consumer's bills.

While it seems unreasonable to suggest, as did one national columnist, that 'conventional rural opinion is now probably, if anything, anti-wind power' (Ridley 1994), the opposition cannot be written off as simply being mistaken, or as just reflecting effective lobbying by a small minority of activists. Even though the opinion polls seem to indicate generally high levels of support (see Box 12.1), there is also clearly genuine local opposition in some areas. The lobbyists could not operate effectively unless there were objections at the grass roots level, and the extensive local press coverage does indicate that there is no shortage of local people who are seriously concerned.

Although NIMBY-type responses seem to predominate, there are also sometimes wider regional preservation and conservation concerns. Certainly, in addition to the issue of visual intrusion, which is a human concern, wind projects might be expected to have some impacts on wildlife and the environment. The issue of bird strikes has not been seen as very significant in the UK (the RSPB has generally backed wind projects) although it has been raised in the USA. Studies have indicated that birds avoid moving objects like wind turbine blades and are much more likely to collide with stationary power grid cables, as happens regularly. Clearly, care must be taken to avoid migration routes, where large numbers of birds are involved. Most other animals seem indifferent to wind turbines. Cows and sheep graze right up to the tower bases. Indeed some seem to welcome them as providing shelter from wind in winter.

Some damage can be done to the local ecosystem during construction, and by the foundations for the concrete bases of the turbine towers, but the wind turbines and their bases can be removed, if necessary, leaving the site more or less unscathed. Wind farms are not allowed in the statutorily defined Areas of Outstanding Natural Beauty (AONBs) and Sites of Special Scientific Interest (SSSIs), although some objectors have complained about wind farms spoiling the view *from* such locations.

Some objectors accept the global argument (e.g. in relation to greenhouse emissions) but claim that wind farms could not help that much, so that the local impact is not justified. Others favour alternative energy options – energy conservation often being seen as a better choice. For some others, the central issue is what they see as 'profiteering' by 'greedy developers',

Box 12.1

Results of some public opinion surveys on wind farms

Kirkby Moor

A study of 250 local residents near the 12-turbine wind farm at Kirkby Moor in Yorkshire was commissioned in February 1994, six months after start-up, by National Wind Power. It revealed that:

- 82 per cent supported the development of wind farms in the area and 84 per cent thought that more energy should be generated from renewable sources;
- 83 per cent were 'not at all concerned' or 'not very concerned' about the noise that they make;
- of those who could see the wind farm from their houses, 77 per cent were 'not at all concerned' or 'not very concerned' about the impact on the landscape.

Taff Ely

A study of 250 local residents near the 20-turbine wind farm at Taff Ely in Wales was commissioned in February 1994, six months after start-up, by East Midlands Electricity. It revealed that:

- only 2 per cent strongly opposed the development of wind farms in the areas;
- 75 per cent said that either they could not think of any disadvantages of wind power or there were no disadvantages;
- noise was not perceived as a problem with only 3 per cent saying they could hear the wind farms from their homes.

Source: Data relayed by the British Wind Energy Association 1994

who make use of the 'extensive subsidies', without, allegedly, being concerned with the impact on the local environment.

On the local proponents' side, strong support for wind farms is seen as part of a positive commitment to the future, with the threat of 'global warming' often being cited, along with the dangers of nuclear power. Many supporters also say they actually like the look of the wind farms. On the other hand, those who do not, obviously feel strongly about it. Thus *Newsweek* (28 March 1994) quoted Sir Bernard Ingham as saying 'people who think they're attractive are aesthetically dead'.

The debate was to some extent set in a more productive vein by the extensive review carried out by the all-party House of Commons Welsh Affairs Select Committee in 1993–4. This took evidence from all the protagonists. The Committee concluded that as long as they were sensibly planned wind farms would be acceptable in Wales, and the Committee in effect rejected some of the more aggressive claims of the objectors (Welsh Affairs Committee 1994).

The subsequent publication of Siting Guidelines by the Friends of the Earth (FoE 1994) and by the British Wind Energy Association (BWEA 1995) also helped clear the air to some extent, as did the publication of further local opinion surveys showing overall support.

However, local objections and national-level lobbying continued, and, as more wind farm proposals emerged, opposition seems to have strengthened. This has led to something of a crisis for the industry and for the government's renewable energy programme.

A crisis of acceptance

As we have seen, most of the initial series of wind farm projects supported under the first NFFO got through and, although objections began to emerge subsequently, 54 MW of the 84 MW of wind capacity contracted for under the second NFFO has been successfully commissioned and eleven wind farm projects supported under the third round of the NFFO obtained planning permission, while only seven were turned down.

However, since then the rate of refusals has increased, so that by the late 1990s nearly 70 per cent of the projects failed to get planning permission (see Figure 12.1). Between September 1991 and December 1993, twelve wind farm proposals went before planning inquiries and nine of these were approved. But between January 1994 and January 1999, only two of the eighteen proposals called in for decision by the minister won approval following a planning inquiry.

Only a proportion of wind farm proposals are called in for planning inquiries, but even so the trend seems clear. Dr Peter Musgrove, from National Wind Power, told *The Times* (9 January 1999): 'Since 1994, planners and inquiry inspectors have been giving progressively less weight to the clean energy benefits of wind farms and progressively more to their negative and subjective assessment of visual impact.'

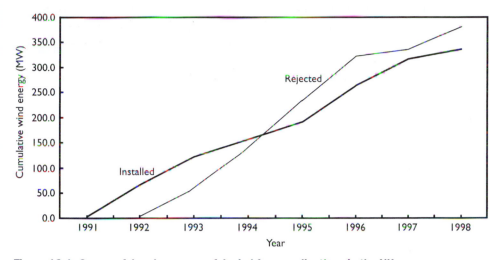

Figure 12.1 *Successful and unsuccessful wind farm applications in the UK*
(*Source:* Royal Society/Royal Academy of Engineering, 'Nuclear energy: the future climate', 1999.
Data from Sinclair Environmental Information Services)

As *The Times* put it, 'unless urgent action is taken many firms will leave Britain for windpower opportunities overseas', and the UK would find it hard to meet the target of obtaining 10 per cent of its electricity from renewables by 2010.

In the next chapter we will look at what might be done to resolve this problem. One approach would be to carry on regardless, possibly adjusting the planning rules to make it easier for wind projects to be passed, on the argument that the opponents were very much a minority and wind power was urgently needed to meet the environmental targets. However, another approach would be to recognise the need to negotiate between the various interest, in terms not too dissimilar from those presented in the model of interactions in Chapter 1, so as to balance costs and benefits more equitably.

Summary points

- The economic constraints in the early NFFO rounds have led to some invasive siting and precipitated some negative local reactions.

- Familiarity generally leads to more acceptance, and overall support was relatively high, but pressure groups can create an impression of widespread objection.

- Sensitive consultation and careful choice of sites are vital if public acceptance is to be obtained.

- In the absence of acceptance, even positive environmental projects can be blocked.

Further reading

The three-volume report of the Welsh Affairs Select Committee on 'Wind Energy' (Session 1993–4 Second Report, HMSO, London, July 1994) gives a good overview of the debate on wind farms in the UK. Volume 1 summarises their conclusions.

The Department of Environment (now DETR) Planning Policy Guidelines on renewables (PPG 22) can be obtained from HMSO, London. ETSU (now known as 'Future Energy Solutions') can also supply various guides for planners and developers.

13 Public acceptance: the need for negotiation

- Local opposition to renewable projects
- The need to negotiate public acceptance
- Developing 'social control' of technology

The impacts of renewables are generally much less than those of conventional energy technologies, but there is still a need to negotiate public acceptance. This chapter reviews the wind farm case study and looks at how the debate has continued, with the focus moving on to attempts to ensure that local communities can have more direct involvement with, and benefits from, projects like this. Local involvement is vital in that, rather than seeing local concerns as a problem, well-informed criticism might also be seen more positively, as an attempt to subject technology to some form of direct social control. After all, one of the alleged benefits of at least some types of renewable energy technology was that it was likely to be more amenable to local democratic control than the preceding large-scale, centralised technologies.

Opposition to technology

Opposition to new technology is nothing new. There are inevitably fears that new technologies will result in major social or environmental dislocations, and sometimes this has proved to be the case. However, the scale of opposition to wind farms has, so far, been trivial by comparison with the opposition that has emerged to nuclear power. The same is true for the other renewables. Even so opposition does exist, and in part this is because renewable energy systems tend to be very visible.

This has been a particular problem for wind turbines and this problem has not been restricted to the UK. For example, during the initial 'wind rush' in California, large numbers of sometimes poorly designed machines were installed near to major highways, so that they were seen regularly by many people. Paul Gipe notes in *Wind Power Comes of Age* that the developers 'could not have picked a worse place to begin than the Altamont Pass' since the meteorological conditions meant that 'for much of the year, and especially during the morning commute, even the best turbines in the best locations stand idle. And all the early trouble-prone turbines were installed immediately adjacent to Interstate 580, one of the busiest highways in the state, with 36 million vehicles passing by per year' (Gipe 1995: 275).

Although, Gipe claims, well-designed machines on good sites might be expected to operate 50–75 per cent of the time, that still meant there would be times when they were becalmed, or out of service for maintenance. The very visible spectacle of inoperative machines suggested to passers-by that they were an inefficient source of power.

However, leaving visual effects such as this aside and assuming that noise problems are avoided, wind turbines have few other impacts compared to some of the other renewables. By contract, waste-into-energy plants involving power production from the combustion of domestic and industrial waste seem to have more potential problems. Some of these projects have met with strong local opposition from residents concerned about toxic emissions and unconvinced by assurances that these would be kept within controlled limits. We reviewed some of the contested issues earlier, in Box 7.4. National-level environmental groups also opposed waste combustion. Greenpeace mounted a very successful 'Ban the Burn' campaign, labelling waste combustion plants as 'cancer factories'. In addition to the issue of emissions, environmental groups like Friends of the Earth felt that energy from waste combustion, as a profitable waste disposal option, would undermine efforts to increase the scale of waste recycling and the development of a more sustainable approach to waste management. For example, given that there are economies of scale with waste combustion plants, large plants are favoured, and there have evidently been occasions when waste has had to be transported long distances, even from other cities, in order to keep the plants fed with fuel.

Interestingly, in this context, in 2002, when refusing permission for an expansion of a waste-to-energy incineration plant at Edmonton in London, Energy Minister Brian Wilson noted that, although the government supported properly sized plants, 'our policy is that statutory recycling targets must be met and that no incineration proposal shall be permitted which will pre-empt recycling or reduce the option for recycling for the future'.

Similarly, there have been objections to some geothermal projects, some of which have resulted in the emission of noxious fumes. In Paris, in 1980, there were objections to the noise during the drilling phase: local residents put up a poster saying 'Oui à la géothermal, non aux nuisances'. More seriously, in Hawaii there were some bitter conflicts over geothermal projects, fuelled in part by local resentment at what was seen by some people as the desecration of their natural heritage.

In the case both of waste and geothermal projects, local opposition has sometimes resulted in their abandonment. This should not be a continuing problem for geothermal plants, at least in terms of emissions, given the use of closed loop technology, with the water being re-injected. However, in the case of waste combustion, public concerns about emissions have been a serious problem in the UK, and that looks like continuing. Many of the waste combustion projects offered support under the Non-Fossil Fuel Obligation (NFFO) could not be completed: only around 72 MW of the 311 MW of municipal and industrial waste combustion capacity supported under the first two NFFO rounds was successfully commissioned. Subsequently, conventional 'mass burn' combustion projects were excluded from the Renewables Obligation – ostensibly, since they were deemed to be commercially viable on an independent basis so that they no longer needed extra support. Instead the UK government called for a focus on what they saw as potentially cleaner waste-processsing technologies – pyrolysis and gasification – which were included as being eligible for support under the Renewables Obligation.

It is not yet clear whether these new energy recovery technologies will win favour with environmentalists – the technologies are as yet relatively underdeveloped. But it is clear that most environmentalists feel that the case against conventional 'mass burn' waste combustion is very strong – it is not seen as a genuine or safe renewable energy source. By contrast, most strongly support wind power. However, as we have seen, some wind farm projects in the UK have had to be abandoned, following local opposition and adverse planning decisions. Indeed, the situation has worsened in recent years. What might be done to resolve this problem?

Possible resolutions: local involvement

As we have seen, there is broad support for wind farms in the UK, and local support actually seems to grow once wind farms have been established. For example, a study on people living near Scotland's first four wind farms indicated that 40 per cent had felt there would be a problem but only 9 per cent reported any. Overall, two-thirds of those asked found 'something they liked' about wind farms (NATTA 2001). Moreover, a poll carried out by the British Market Research Bureau for the RSPB in 2001, found that only 3 per cent of those asked objected to wind farms. However, there is also some strong local opposition and, even if it represents a minority, it seems to be slowing the rate of deployment dramatically.

In looking for a solution to this problem, we have to first recognise that not all wind projects are necessarily well thought out. Certainly, there are a variety of practical steps that could be taken to improve the situation, such as better location of wind farms and the use of waste land or old industrial sites. Technological developments may change the situation. The introduction of more efficient variable-speed wind turbines, which can operate effectively with lower wind speeds, could reduce pressure on upland sites. Costs have also dropped dramatically since the early days, to one-third of the cost of wind projects established in 1990. That makes it more economically viable to choose less windy sites.

However, to ensure that these technical and economic improvements translate into environmental/locational improvements will probably require something more coherent than the UK's current rather *ad hoc* approach to planning in relation to wind projects, which operates on a site-by-site basis. Obviously, it is sensible to assess each project on its merits, with local issues and concerns being taken into account. This can lead to expensive planning delays and one solution would be to develop some form of zoning, with areas that are suitable for development being identified in advance, as in Denmark. Something like this is in hand in England with the development of a regionally based planning framework and local councils setting targets for the amount of renewable energy to be obtained in their regions (DETR 2000). There is also an emphasis towards streamlining the planning system to avoid long delays and provision being made for Parliament rather than local inquiries to determine some major infrastructure projects. However, if that led to any weakening of planning control it would most likely be seen as reprehensible by the majority of environmental groups, even those in favour of wind, since it might open up the way for less desirable projects.

Nonetheless, local opposition is likely to persist in many places unless a way is found to address the feeling of powerlessness that some opponents evidently feel in the face of wind projects imposed on their communities. Most of the wind projects are owned and controlled, one way or another, by a few large companies, and this has sometimes been a source of resentment. For example, one of the issues raised during the local campaign against the 39-turbine wind farm given planning permission for a site at Cefn Croes in mid-Wales was that it would will not benefit local people and was being 'imposed on the community by Enron Wind, a subsidiary of the shamed US multinational' (*Guardian*, 20 February 2002).

By contrast, opposition to wind farms has been far less apparent in Denmark and Germany where, as we have already noted, the majority of wind projects are owned by local people, who share in the profits. The growth in privately owned as opposed to utility-owned projects in Denmark has been quite striking (see Figure 13.1). Around 80 per cent of 300 MW of new wind capacity installed in 1997 involved projects owned by individuals or local wind co-ops, called Wind Guilds. There were changes in the subsidy system in the mid-1990s which slowed the rate of growth of locally owned projects, but even so by 2000, 75 per cent of the 900 MW then installed were owned by local co-ops or individuals, in roughly equal numbers, and half of the large 40-MW offshore wind farms installed in 2001 2 km off the coast from Copenhagen are owned by 9,000 local co-op members. Overall, more than 100,000 families own shares in local wind projects in Denmark – out of a population of 5 million. Local ownership via community-based co-ops (known as Windmill Associations) is also popular in the Netherlands. Similarly, in Germany, local interests dominate ownership, with around two-thirds of the project being

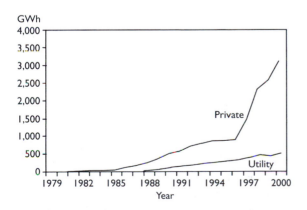

Figure 13.1 *Growth of local ownership of wind projects in Denmark*

(*Source:* Data from Paul Gipe)

owned/operated by local farmers, home-owners and small businesses – and, as we have seen, by 2002 Germany had installed around 10,000 MW of wind capacity, compared with the UK's 500 MW. While in the UK some people have been clamouring for wind projects to be halted, in much of the rest of Europe many people seem to be clamouring to be able to benefit economically.

If wind projects involve some form of local control and local economic benefits, the local impacts may seem to be judged more favourably. Danish enthusiasts for local ownership of wind projects often repeat the old Danish proverb 'your own pigs don't smell'. Put more positively, to the extent that the problem is local residents' sense of powerlessness in the face of what they perceive to be projects imposed on them by outsiders, who profit from them, then the solution is for members of the local community to become directly involved, via ownership. That can lead to direct economic benefits as well as to more indirect gains, such as local economic regeneration. For example, farmers, who are finding it hard to survive economically by producing food, might diversify into wind power and other renewable projects, thus strengthening the local economy (Elliott and Toke 2000).

So far, there is only one wind co-op in the UK, Baywind in Cumbria, although some other initiatives are underway. One group in Wales, Awel Aman Tawe, has taken pains to consult widely with local people, and in 2001 it organised a public referendum in the area to decide whether a wind co-op project should go ahead (the result was 57.5 per cent for, 42.5 per cent against).

Of course, even with conventional commercial projects, there are some local benefits, in terms of employment during construction and tourism subsequently: some sites have set up visitors centres which have proved popular tourist attractions, injecting cash into the local service economy. Local farmers can also benefit, by renting out fields for use by wind farm developers (rentals of £1,000 to £2,000 per acre or per turbine are typical), although this can sometimes lead to local conflicts and resentment if residents feel they are adversely affected by the wind farm.

In some cases wind farm developers have offered packages of 'community support measures'. For example, National Wind Power provided £100,000 for a charitable trust to be used to support local schools, colleges, students, apprentices and training schemes in the area around its wind farm at Bryntitli in mid-Wales. It has also provided a £5,000 p.a. fund to support local environmental improvements in the area of the wind farm, and has set up a community fund that will receive £5,000 p.a. for the benefit of local inhabitants (NATTA 1994c). Schemes like this might give local communities a vested interest in the project, or failing that, might be seen as compensation for any disbenefits. Equally, they might be seen, by opponents of the project, as a form of bribery.

In principle, local involvement could range from the purely token, right through to full-scale local ownership. However, in the UK economic environment, the latter seems a little unlikely, for the moment, on any significant scale. More likely, there will be continued conventional corporate developments, softened perhaps with some attempts at local-level involvement. In which case there are also likely to be continued allegations of corporate insensitivity and 'profiteering', with the wind programme being seen, at least by objectors, as being driven by profit-oriented commercial interests.

As has already been indicated, this may be unfair, not simply because the profit margins are in fact very tight, but because many of the developers have strong environmental commitments. But, from the grass roots NIMBY viewpoint, this may not be convincing. Sensitive local consultation, well in advance, is an obvious priority, but that takes time.

Probably the key issue will therefore be the pace of any future programme – as well, of course, as its scale. If a large and rapidly expanding programme is seen as being driven purely by profit concerns, then opposition is bound to grow. If, however, a commitment to wider social and environmental goals can be convincingly displayed, then perhaps more support will be forthcoming: many people are, after all, concerned about the greenhouse effect and acid rain. In order to engage with that concern, the wind power lobby will have to win the argument that wind power can make a significant contribution.

Another form of involvement, albeit rather diffuse, can occur if consumers contract with one of the dozen or more green power retail schemes that are on offer. As was mentioned in Chapter 4, in these schemes, electricity suppliers promise to match the power consumers use with power bought in from renewable sources. In some schemes a small surcharge is made. By 2002 around 60,000 people had signed up in the UK, and the number is expected to grow significantly, as has happened elsewhere (for example, by 2002 there were 324,000 green power subscribers in Germany and over 1,300,000 in the Netherlands). Some of the schemes use wind power, and participating consumers are able to identify the sites used. This may not be ownership, and it may not be a local plant, but it does make some sort of link between what comes out of the socket in the wall and the source of the power. Moreover, in some schemes, consumers actually donate a surcharge direct to a fund or trust that invests in specific renewable energy projects, which are usually local. So, in that case, there is an even more direct sense of local involvement.

There are also other changes that are occurring in the electricity market, which might provide an opportunity to help improve the situation. Before liberalisation of the market, everyone in the country paid the same for electricity regardless of where they lived. Now consumers may have to pay different amounts depending on where they live and from whom they buy their power. Before liberalisation, people in remote areas were, in effect, subsidised by those who lived nearer to power stations – since there are power losses associated with long-distance transmission. But, if power is produced locally, it is clearly valuable to use it to meet local energy needs. In which case, should not people who live near wind farms be charged less than those at a distance? That would compensate directly for any loss of local amenity. However, this idea opens up the possibility that other groups in society might seek to claim compensation, for example, for living next to a nuclear plant. But it would establish some sort of link between costs and benefits.

Conclusion

'I suspect that future generations will look back at controversy over wind farm sites and wonder what all the fuss was about. They will be part of our landscape and seascape and we will be pleased to have them. Although it is important to take local wishes into account, we should not allow a NIMBYist stance to impose a veto on onshore wind farm developments.' So said Peter Hain, then Energy Minister, during a House of Commons debate on Renewable Energy, in 2001 (Hain 2001).

It is certainly interesting that there were tens of thousands of traditional windmills dotted around the UK until the eighteenth century and the survivors are now cherished as beautiful. However, it is perhaps too easy to depict objectors to modern wind turbines as simply responding to NIMBY sentiments. The debate can also be seen as reflecting an important concern for environmental protection. As an opponent of a Scottish wind farm put it, 'We are the custodians of this landscape for future generations and as such it is our duty to fight this alien intrusion.'

One of the fundamental issues that emerges from the wind case study relates to people's sense of a lack of power to protect their interests, which may include altruistic environmental concerns. However, in some cases, the development of environmental sensitivity can lead to some contradictions when it comes to choosing the focus for specific campaigns. For example, if you are concerned to protect your local environment, it can seem to be far more effective to oppose a local wind farm than to campaign to halt global climate change. The former can be seen as an immediate and visible threat; the latter is remote, longer term and, in any case, seemingly beyond your influence.

Some objectors have tried to adopt a more positive approach by pointing to alternative possibilities – for example, offshore wind farms. As we have seen, the UK's offshore wind resource is very large and the environmental impacts associated with tapping them are generally low. But, given the UK's renewable energy target of 10 per cent by 2010, offshore wind is likely to be additional to, rather than instead of, on-land siting. Moreover, offshore siting is more expensive than onshore wind, primarily due to the need to transmit power by undersea cable to the shore, and, unless offshore wind projects are sited well out to sea (thus increasing cost further), there could still be objections about visual intrusion. Indeed in 2002, there were strong local objections to the offshore wind farm proposed for a sandbank five miles out from the Scottish side of the Solway Firth, and there was a local campaign against the project proposed off the coast of south Wales near Porthcawl.

Energy conservation is obviously another alternative, but once the opportunities for cheap and easy savings have been exhausted, the cost effectiveness of energy-saving measures is not much different, in terms of carbon emissions avoided per pound spent, than that obtained from wind power projects. Moreover, if the climate change problem is as serious as many scientists think and we wish to avoid the nuclear option, then we will need all the energy conservation and all the renewable sources we can reasonably muster – it is not a matter of either one or the other. This conclusion becomes even more stark if we assume, as seems likely, that, even given a major commitment to energy conservation, overall demand for power will continue to rise, given the ever growing lifestyle expectations of our consumerist society.

Of course, this situation is not unchangeable. As we will be discussing in the next part of the book, we may be able to move to more sustainable forms of consumption. However, that could require major social changes and, if it were to be in any way equitable, there would be a need for a major exercise in negotiation among the various competing interest groups in society. Be that as it may, for the moment it seems that most people cannot accept life without more and more energy-using devices, from dishwashers to tumble dryers. In which case the point has to be made that the energy will have to come from somewhere. If we want to avoid the problems of climate change and other environmental problems associated with the conventional energy technologies, including nuclear power, then we will have to develop the renewable alternatives.

Negotiating the future

Given this situation, it would seem vital to engage all parties in a constructive debate over how we make use of renewables and, in particular, how to deploy them with the minimum impacts. As we have seen, the wind farm debate in the UK has become quite strongly polarised. Although clearly facing problems, the pro-wind lobby remains confident that as more people see real wind farms, rather than reading perhaps misleading press reports about them, support will grow. At the same time, the anti-wind lobby has become increasingly confident, given its success at gaining media access and political attention.

Perhaps, rather than confrontation, the debate should lead to some form of negotiation on the extent to which wind farms can be introduced in the UK countryside. It is relatively easy to depict objectors as simply responding to NIMBY sentiments. But they can also reflect an important concern for environmental protection which must not be lost in the wider strategic debate over the role of renewables in the UK energy context, for example, concerning which technologies should be emphasised and how rapidly they should be developed.

Part of the debate over how to develop these new energy options concerns the relative distribution of costs and benefits of new energy supply technologies like wind turbines amongst and between, for example, urban and rural dwellers. Inevitably, since the bulk of the population in the UK lives in urban areas they get the energy benefits while rural areas have to deal with any negative impacts. We have seen that there could be ways in which rural people near wind farms could be compensated for any local costs or could share in the benefits. As we have also seen, in countries like Denmark and Germany, rural residents are now actively involved in generating and selling power.

Unless measures are taken to bring about similar changes in the UK, it could be that we will see even more local opposition to wind power, and possibly to other renewable energy projects impacting on rural areas, such as growing energy crops.

More positively, a constructive public debate on wind farms could lay the basis for the sound development of the other renewables. A debate is important since the renewables represent essentially a new form of energy technology, with new types of impact. Whereas previously the emphasis has been on concentrated energy sources and centralised power plants, the trend seems to be towards the more decentralised use of diffuse natural energy flows and sources. The emphasis would thus be on natural processes occurring in 'real time', not with inherited wealth from stored fossil or fissile energy. That opens up a whole new range of planning and land use issues, which have only just begun to be discussed.

As yet, there are few techniques of analysis to help us assess the relative importance of any social and environmental impacts. The use of cost–benefit analysis, in which an attempt is made to give economic values to costs and benefits, seems unlikely to provide anything more than a very partial means of assessing renewables. New techniques seem to be necessary – for example, reflecting the impact of extracting energy from diffuse natural energy flows and perhaps following the approach being developed by Clarke as discussed in Chapter 3. There is also a need for more detailed studies of how public awareness and understanding develops and changes. And there is a need for new methods of social negotiation to resolve the inevitable social conflicts.

One practical possibility is to develop on the UK Environmental Council's Environmental Resolve initiative, which involves training workshops on conflict resolution techniques, designed for planners, developers, environmental groups and so on. The British Wind Energy Association made use of a 'consensus-building' approach, involving discussions and negotiations with all the key parties, including groups opposed to wind power, as part of the process of drawing up its Best Practice planning guidelines (BWEA 1995). Consensus-building approaches are becoming increasingly common in the renewable energy planning field (Hyam 1995). At the same time there is also a need for more general social mechanisms for consultation and conflict resolution.

There is also a need for more information on the issues to be disseminated to the general public, in order to improve the quality of the debate. In Denmark, the national referendum on whether to embark on a nuclear power programme was preceded by a major national education effort, via local meetings, seminars and so on. Environmental education generally seems to be a vital prerequisite for informed debate on how best to develop sustainable energy technologies.

While some renewable energy enthusiasts have expressed concern at the tone of some of the current debate in the UK over the merits of wind farms, in general public debate should be welcomed, as long as it is as well informed as possible. Renewable energy technology is meant to be both socially and environmentally acceptable, and there is a need for public debate over specific projects and over longer term development patterns. That is part of the process of bringing technology under more direct social control. Of course, as with all exercises in democracy, there can be costs, not least in terms of delays. To try to limit these, there may be a need to develop new ways of conflict resolution, via an extension of public consultation and participation in decision making and planning processes.

In principle, the development of renewable energy should be more amenable to local social control since many of the technologies are on a relatively small scale and their nature, function and likely impacts are relatively easy to understand. We will be returning to look at some of the specific issues that might form part of the debate over how best to develop renewables in Part 4.

To summarise, our case study of the pattern of initial opinion formation highlights the problem of bias, prejudgement and conflicting social and environmental priorities. These problems are not unique to wind power, but one would hope that they may be somewhat less intractable, given the 'transparent' nature of the technology and its impacts. Box 13.1 pulls together some ideas for a way of dealing with wind power decisions that might be more effective than simple confrontation on a site-by-site basis. What is needed is a constructive, well-informed debate on how best to develop renewable energy technologies like wind farms, as part of a process of bringing technology under social control. That would seem to be a vital prerequisite for any attempt to move towards a more sustainable energy system.

Box 13.1

Wind farm decisions, issues and processes

1 One would hope that the developers would carry out extensive local consultation well in advance to explain the proposal and try to avoid any conflicts, and possibly modify the plan in response. Ideally local projects should fit into an agreed plan for the regional development of renewables.

2 If the objections are still widespread, the proposal would be likely to be subjected to scrutiny via a public planning inquiry, assuming that the local planning system had not already refused planning permission. The project may then be abandoned.

3 Assuming the objections are only from a minority and are deemed less significant, the authorities may seek to push ahead regardless, although this risks a confrontation.

4 A more creative approach would be to seek to include an element of local ownership in the proposal, so that local people would share in the economic benefits of the project, to compensate for any environmental/local amenity costs. Direct forms of compensation may also be considered, but can seem like bribery.

5 Locally, the outcome of the debate over wind projects will reflect the relative power of the two sides, neither of which is limitless: the objectors will probably not want to push opposition to the extreme (direct action, sabotage) and the authorities will not want to create undue disaffection by draconian actions.

6 The debate, locally and nationally, might usefully be set in global terms – renewable energy technologies like wind farms are part of the solution to the global problem of climate change, and any local impacts could be assessed in this context, with the final issue being, if not wind then in what other way are we to meet our energy needs without damaging the global environment?

Summary points

- Local planning and land use issues may well determine the scale and pace of renewable energy development.

- Some objections may have been based on misinformation.

- Not all objections are wrong: there is a need for careful assessment and negotiation to avoid problems.

- Well-informed criticisms might be seen as part of an attempt to bring technology under social control and part of the process of moving towards a sustainable society.

Further reading

For updates on the wind farm debate in the UK see the series of reports produced by NATTA, the most recent (volume V) being 'Windpower in the UK: into the millennium' (2002). They are compilations from the coverage in the NATTA newsletter *RENEW* (see http://eeru.open.ac.uk/natta/rol.html). This also covers the parallel debates on other renewables.

The British Wind Energy Association's web-site carries useful reports on developments: http://www.bwea.com.

For the anti-wind perspective see the Country Guardian web-site: http://ourworld.compuserve.com/homepages/windfarms/.

For an example of a local campaign against a wind farm see: http://www.cefncroes.org.uk.

For an example of a campaign to develop a local wind co-op in Wales see the Awel Aman Tawe (AAT) web-site: http://www.awelamantawe.co.uk/.

The story of the Danish wind co-ops is told at: http://www.windpower.dk/articles/coop.htm.

 # Sustainable society

In Part 4 the emphasis moves from specific sustainable energy technologies and their problems and on to the wider issue of the prospects for sustainable development generally. It asks whether technical fixes will suffice or whether there will be a need for more radical changes. Will the use of sustainable energy technologies allow economic growth to continue, or are there ultimate environmental limits that will constrain human aspirations and expectations?

By way of conclusion, Part 4 also looks at some of the ways in which the conflicting interests of people and planet might be resolved, and at the idea of 'thinking globally and acting locally'.

14 Sustainable development

- The limits of technical fixes
- The need for social change
- Incentives for change
- Strategic choices for the future

Given that we live on a small planet with finite natural resources and a fragile ecosystem, environmental sustainability may be incompatible with continued economic growth, at least of the current sort. But, as this chapter shows, not everyone agrees with this proposition: the optimists believe that human ingenuity and technical fixes will suffice; the pessimists believe that more radical technical, and perhaps social, changes may have to be made. Radical changes may appear threatening to those whose livelihood is linked to the present arrangements but, as we shall see, there may also be some benefits.

Strategic issues for sustainable development

The key message of our analysis so far is that technology presents problems, but also possibly some solutions, even if those solutions can have problems of their own. These problems require a process of social negotiation – as part of the larger process of moving towards a sustainable future.

Many strategic and tactical issues have emerged from our discussion so far. Some are essentially 'technical', although they may have major social implications; for example, should renewables be developed on a local, decentralised basis or must they be integrated together on a larger scale? We will be looking at this sort of issue in Chapter 16.

Our focus in this chapter is on strategic issues that take us beyond the technical and raise wider social and political questions, concerning the overall direction not just of technology but of society.

Following on from the criteria developed in Chapter 3 and the discussions in subsequent chapters, the major strategic questions for the future in the energy field would seem to be as follows: to what extent can renewables and conservation help us move towards environmental sustainability, and how fast can and should renewables, and other 'green' technologies, be developed and deployed to this end?

These may appear, initially, to be relatively straightforward 'technical' questions, concerning the ability of technology to deliver the required amounts of energy. However, they raise broader issues, for example, concerning exactly what is meant by 'sustainability'. Perhaps the hardest question concerns the relationship of sustainability to economic growth. Renewables and conservation may be able to sustain economic growth up to a point, but is continued economic growth of the current sort viable or even desirable? A linked, but slightly more tractable question is, will technical fixes suffice or will there also be a need for more general social change?

Technical fixes succeed

From the purely 'technical fix' point of view, a move towards sustainability, in terms of significantly reduced levels of resource use and resultant emissions, seems likely to be technically feasible. Pollution can be cut dramatically by 'end-of-pipe' measures as well as by the introduction of more radical 'clean technology'. A lot has already been done, given the ever increasing pollution controls and environmental regulations that legislators in Europe, the USA and elsewhere have introduced.

Faced with the expectation of continuing progress like this, some people think that human ingenuity can solve just about all our environmental problems. Technology, and technical fixes, can, they say, provide the answers. One of the most famous adherents to this view was Hermann Kahn, the US futurologist, who was critical of what he depicted as the pessimism of the environmentalists. While some environmentalists in the 1970s were drawing up doom-laden scenarios of imminent energy shortages and environmental collapse, Kahn and his followers pointed to technological breakthroughs that could resolve some of these problems. In the short term, Kahn was proved right: the expected crisis went away. In the environmentalists' view, however, it had only been postponed, not resolved. Kahn was unconvinced. His last book, *The Resourceful Earth*, summed up his views: the planet was immensely rich and mankind was clever and could develop technology to resolve most problems (Kahn and Simon 1985).

The technical fix position adopted by Kahn and other similar optimists, is sometimes associated with what has been called the 'cornucopian' view, named after the legendary Greek goddess of plenty. The cornucopians hold that abundance is there for the taking since, unlike other animals, human beings have the intelligence to create technologies to meet their ever-expanding needs and to do so in ways that will not undermine their continued enjoyment of life and the environment. At the extreme, this view can be used to justify an assertion that environmental problems are not very important. These problems can be 'fixed' and there is no need for heavy intervention in economic affairs: certainly nothing should be allowed to slow economic growth. Indeed it was growth that could provide support for the development of new cleaner technologies, should they be needed.

This sort of analysis is often heard on the political right where it is held that there must be no interference with free market competition. The approach has been well represented by Wilfred Beckerman's *Small Is Stupid*, which mounts a serious attack on current green thinking (Beckerman 1995). A wide range of similarly 'contrarian' books have emerged since then, perhaps the most influential being Bjorn Lomborg's *The Skeptical Environmentalist* (Lomborg 2001), which argues basically that environmentalists are prone to scaremongering and that in reality 'things are getting better'. The view that climate change does not warrant major responses, other than adaption, has also become increasingly popular among conservative contrarians, some of whom clearly feel that governments should not be undermining free market competition by subsidising the development of renewable energy (CATO 2002).

This chapter is not the place to enter into what is obviously an ideological debate. Suffice it to say that more liberal-minded critics feel that some form of intervention is vital. The real debate is over the extent of the intervention and how precisely it should be carried out. Modern liberal-minded environmentalists argue that if the market is given the right signals it can be used to bring about appropriate behaviour by investors and consumers. Markets can be shaped by a whole range of fiscal measures such as taxes, and by state subsidies, grants, regulations and controls, in order to prioritise more sustainable developments.

The UK government seems to be taking this challenge seriously. Thus DTI Minister Brian Wilson commented at the launch of the Sustainable Technology Programme in 2002, 'If we are to truly meet the challenges of sustainable development, then the role of Government in the modern world must be to promote innovation in a way that encourages industry to think in terms of long-term sustainability, rather than solely aiming for short-term economic gain' (Wilson 2002).

Increasingly, the substance of debates over modern politics and economic policy in the developed countries concerns exactly what mechanisms should be used and how they should be applied.

However the key strategic questions still remain. Whatever the mechanism used to stimulate them, can technical fixes really solve our environmental problems, or will there also have to be social changes? And even if technical fixes can resolve most of our current problems, can growth continue indefinitely? Can we continue to find ways to improve the efficiency with which we use resources, in order to avoid the problems associated with using more of them?

Prime Minister Tony Blair seemed to recognise some of these issues when he said in a speech to the CBI/GreenAlliance meeting in October 2000,

> the central theme of our approach is a more productive use of environmental resources. It is clear that if we are to continue to grow, and share the benefits of that growth; we must reduce the impact of growth on the environment. Some commentators estimate that we'll need a tenfold increase in the efficiency with which we use resources by 2050 only to stand still.

Beyond technical fixes

To try to get to grips with these question, let us go back to the basic issue of energy supplies. As we have seen, energy wastage can be reduced dramatically in most sectors of use (domestic, industrial, etc.) by the introduction of energy conservation techniques and the use of energy-efficient devices, products and production systems, making it easier for renewables to supply the remaining energy requirements. Savings of at least 50 per cent and perhaps up to 70 per cent or more are seen as technically possible. As Figure 9.1 illustrates, in principle renewables could then meet around 50 per cent of global energy requirements by 2060 or so.

The study produced for Greenpeace mentioned earlier looked even further ahead (see Figure 14.1). Greenpeace estimated that renewables could supply nearly 100 per cent by 2100, given proper attention to energy conservation. So there is no 'doom and gloom' here. Indeed Greenpeace claimed that this could be achieved without their being a need to curtail economic growth. In their scenario, primary energy consumption increases dramatically. This is based on the assumption that the overall world gross national product continues to increase, although an attempt was made to factor in some redistribution between countries in the North and South (Greenpeace 1993).

Long-range scenarios like this are useful as a way of mapping out not what will happen, but what is technically feasible. The Greenpeace scenario is based on some fairly optimistic assumptions about the successful development and deployment of new renewable energy technologies. For example, it assumes fairly widespread use of photovoltaics, hydrogen-powered fuel cells and biofuels. But it does seem reasonably realistic, especially since it does not exhaust all the technical possibilities; for example, it does not assume any contributions from wave power or tidal power.

The real issue of course, apart from whether the specific technical options can deliver, is the question of implementation. Greenpeace notes that the

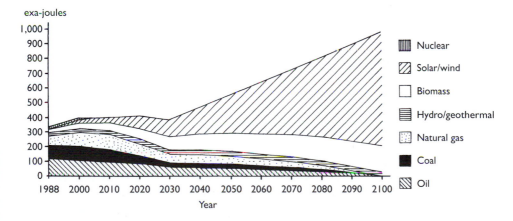

Figure 14.1 *Greenpeace's fossil-free global energy scenario*
(*Source:* Greenpeace, 1993)

key factor in bringing this about was the political will to make the transition, with the implication that there would have to be serious commitments from and intervention by governments.

Obviously, this is a radical scenario – and no one as yet has adopted a programme anything like it. But there are some signs of interest in making a start. For example, in 2000 the UK's influential Royal Commission on Environmental Pollution proposed that the UK should aim to cut its greenhouse gas emissions by 60 per cent by 2050, and some other countries seem to have concluded that global environmental problems like global warming warrant an even more radical approach (Royal Commission on Environmental Pollution 2000). Given that the advanced industrial countries benefited in the past from using the world's fossil fuel resources, it has been argued that they should make even larger reductions in emissions, of up to 80 or 90 per cent, so as to allow the global average to be 60 per cent.

It might be possible to achieve this level of reduction via technical fixes, without having to reduce the actual level of energy consumption at the point of use. For example, the Greenpeace scenario assumes increased global end use of energy as well as increased primary energy use, while still achieving dramatic reductions in carbon emissions.

However, the Greenpeace scenario may be optimistic: it may be that the various technical fixes will not allow overall levels of growth in energy use to increase in this way, at least not everywhere in the world. As we noted, Greenpeace assumed a degree of economic redistribution between the North and the South and, presumably, lower rates of growth in energy use in the advanced countries, but this may be optimistic. Even given technical fixes and improved efficiency, there could be a need for more than just a slowing of growth rates to achieve a sustainable balance. There might be a need for overall reductions in the absolute levels of material consumption, particularly in the advanced countries. At the very least there could be a need for overall patterns and levels of consumption to change, with perhaps a move away from the Western emphasis on 'quantity of material consumption' to 'quality of life'.

The idea that consumption might be voluntarily limited is gaining ground. There are already 'sustainable consumption' initiatives in Scandinavia, where it is sometimes argued that consumption levels are, in any case, reaching saturation levels (Norgard *et al.* 1994) and there has been some debate in Scandinavia and in the USA over the idea of 'voluntary simplicity', that is, consciously choosing a less consumerist approach to life. Similar views have emerged elsewhere. For example, in Australia there is a strong grass roots environmental movement, the radical wing of which seems convinced that technical fixes will not be enough, and that there will also be a need for major social changes (Trainer 1995).

Some greens believe that social and technological change is not only environmentally necessary, but is also morally and socially desirable, given the vast imbalance of affluence around the world. This amounts to a call for a new set of environmental and global values. In some cases this can be taken to imply a radical change in lifestyle, new ways of thinking and more co-operative ways of organising to meet needs. For example, the 'deep ecology' view, which began to emerge in the mid-1980s, stresses the need for human beings to get back in touch with nature, and for a new emphasis on the spiritual as opposed to the material (Deval 1988). This approach is often two-edged. On one hand, the deep ecologists claim that living in this way could actually be more fulfilling, but, on the other, they insist that human beings have no right to live in any other way.

An end to growth?

Underlying many of the 'deep ecology' critiques of contemporary industrial society that have emerged from the radical end of the green movement is the idea that economic growth, at least of the type experienced so far, cannot be continued. For example, it is sometimes argued by deep ecologists that even if massive energy conservation programmes are introduced, toxic emissions cleaned up and a switch to renewables achieved, the planet's ecosystem will still be overwhelmed if the ever-growing world population attempts to attain Western levels of material affluence. The only alternative, in this view, is to move towards more sustainable lifestyle patterns, based on drastically reduced levels of consumption: essentially 'living better with less'.

This is not a new idea: the idea of 'voluntary simplicity' as a response to environmental problems has been around since the 1970s (Elgin and Mitchel 1977). However, it is based on a prognosis that Kahn and the cornucopians claim is unduly pessimistic.

There is no way in which this debate between optimists and pessimists can be easily resolved, at least not within the context of this book. The debate is usually couched in technical and economic terms, with much use of statistical prediction of trends. Quantitative analysis can certainly help to try to identify absolute limits (e.g. in relation to toxic emissions or greenhouse gas concentrations) and statistical patterns (e.g. in population growth or increases in economic productivity). We have already mentioned the idea of an overall energy limit in Chapter 3, and in the next chapter we shall be looking at some of the overall constraints on the pattern of global energy use and industrial development.

In reality, however, much depends on assumptions and beliefs concerning what is possible and what is desirable – and these are not always susceptible to objective analysis. Even some of the terms and concepts used can present problems. This is nowhere more clear than in the way the term 'sustainability' has been interpreted and used. For radical greens its meaning is clear: the long-term survival of the planetary ecosystem, which many hold is incompatible with continued economic and industrial growth. At the same time optimists sometimes talk of 'sustainable growth' or even 'sustained growth', and the term sustainability has become not an absolute but more of an adjective, describing an attempt to move in the direction of more environmentally compatible approaches.

While technical analysis may help to reduce some of the uncertainty, in the end the debate over sustainability and the appropriate means of attaining it would seem in essence to be a political and ideological debate, concerning, for example, conflicting social priorities, generalised environmental ethics and vested economic interests. If this is the case, in practice the debate can only really be resolved by political means – by the messy process of reaching agreements, among people with differing views, interests and amounts of influence, on what should be done.

If the pessimists are right, there would be a need for what could be perceived as painful social changes, accompanied by some hesitancy. For the cornucopian optimists are not the only ones who would find the proposition that lifestyle patterns and consumption levels will have to change unwelcome. Many people in the advanced countries would no doubt see any change in their material standard of living as a serious imposition and will resist them. Many people are happy to engage in recycling and some may purchase environmentally friendly products, but many would no doubt perceive actual net reductions in material consumption as a threat to their standard of living and as an unwanted return to frugality. A shift away from consumerism, and any suggestion of lowered economic growth rates, would also no doubt be resisted by industrial and commercial interests, who might fear that profits would be undermined.

We are back to the conflicts outlined in the model of interests presented in Chapter 1. In this case consumers and producers share a perceived common interest in continued economic growth. This is as true in the developing world as it is in the developed countries: people in the developing world would like access to similar benefits as those enjoyed in the developed countries. It is therefore not surprising that technical fixes are popular: they seem to offer the possibility of the benefits of continued growth without the environmental problems.

While there is no shortage of evidence that there is significant generalised public support for more environmental protection, even if this costs more, whether these generalised concerns could be translated into acceptance of actual reductions in consumption levels remains to be tested (DTI 2002). Voluntary shifts in consumption patterns may occur, as people become more aware of environmental problems, but shifts in attitude and behaviour are more likely to occur if there are some clear benefits. It may thus be that consumers are likely to need positive incentives before they will make major lifestyle changes.

Incentives for change

It is hard for individuals to accept lifestyle changes and it seems likely to be even harder for nations to adopt new approaches that may go against short-term vested interests. As the slow progress on responding positively to the threat of global warming indicates, there is still some way to go before the global community will be able to respond to the serious global environmental challenges that many feel lie ahead. Vested interests in the continued use of existing energy sources still dominate; for example, some of the oil-producing countries have resisted attempts to develop international obligations concerning reductions in carbon dioxide emissions.

In order to try to break out of this sort of impasse there is clearly a need for concerted efforts at negotiation and commitment to action at all levels, stressing communality of interests and trying to resolve sectional differences. After all, in the longer term, all the world's people share a common interest in the health of the planet. A shift to a more sustainable approach will involve significant restructuring of industrial activities, but, as we shall see, there may actually be some longer term commercial advantages. More generally, any disbenefits associated with a shift to sustainability should be set against the overall shared benefits. This, of course, is not always easy, because of the conflicts of interest that already exist between the various vested interests within any particular country and the competing interests of countries themselves.

Reaching international agreements on change is likely to prove particularly difficult, especially if the implication is that economic growth and competitiveness will be affected. Vested interests in the status quo remain strong. For many radically minded people, it is unlikely that agreements can be reached when the world economy is dominated by powerful multinational companies who are seen as having no interest in the fate of individual countries. In the same way, it is also often seen as unlikely that agreements can be expected between the North and South when many of the developing countries are trapped in debt relationships with the rich countries.

From this viewpoint, there will have to be major social, economic and political changes before there can be an effective international response to global environmental problems. At the very least, international economic activities must be subject to some form of control to ensure fairer trading relationships and reduce exploitation of both people and the planet. Obviously, this opens up issues that are well beyond the remit of this book. We will be looking at the overall prospects for economic expansion worldwide in the next chapter, where we also discuss the need for technology transfer and aid programmes, but for the moment suffice it to say that there is a clear need for changed priorities, not least by multinational companies.

Of course, it could be that the major companies of the world may introduce some changes to the way in which they operate, not so much on an altruistic basis but in order to maintain their longer term commercial survival. On this view, since the world's environmental problems will not disappear, and pressures to clean up are likely to increase, companies and countries that invest now in 'clean green technology' will be better placed competitively in the future. Of course, some companies will be able to ignore these problems in the short term and when and if problems emerge, they will move on to other activities where constraints are less. But in the longer term there will be no escape from the environmental and resource limits and from competition by companies who have made the switch to more sustainable approaches.

Some companies have already embarked on quite radical clean-up programmes, and this trend is likely to increase. Indeed it has been suggested that, given the prospects for economic growth and the need to stay ahead of the competition, the global multinational corporation may be one of the key agencies in bringing about a shift to more sustainable approaches (Wallace 1996). Certainly, the market for pollution abatement and environmental clean-up technology is growing: it had reached £130 billion worldwide by 1990 and was over £250 billion by 2000. Similar growth rates might be expected in the more radical 'green products/clean technology' field and in the renewable energy field – one 1996 estimate by the World Energy Council suggested that the latter might grow to $1.4 billion worldwide by 2020.

Some politicians have talked in terms of a 'green industrial revolution' opening up. At a CBI–Greenpeace Business Conference in October 2000, Stephen Byers, then Secretary of State for Trade and Industry, noted that

> the global market for environmental goods and services is currently estimated at 335 billion dollars – comparable with the world markets for either pharmaceuticals or aerospace – and is forecast to grow to $640 billion by 2010. This is why we must aim to be among the front runners in the green industrial revolution.

He went on to describe a vision of prosperity through environmental modernisation:

> We can not continue to pollute the environment and consume resources in the way we have over the last two hundred years. We all know that is unsustainable. We therefore need demanding, long-term objectives and goals to improve our productive use of resources, and to cut waste and pollution. That is not a threat to business, to growth, to prosperity. It is the key to our future prosperity. This is a fact that many businesses already recognise.

Some people thus look forward optimistically to 'win-win' outcomes – commercial advantages combining with environmental improvements, with there being no losers. While it is a welcome change from the view that responding to environmental problems is 'all pain and no gain', it seems likely that there will inevitably be some pain and some losers, at least in the short term, as adjustments are made. In the short term, for example, some companies may not be able to make the transition, and some renewables may be more expensive, so that consumers may have to pay more. In this situation, there are bound to be some commercial casualties.

In the longer term, however, the new sustainable energy technologies should be able to expand to meet energy needs at reasonable costs, without imposing major environmental costs. The net cost to society could well be less – when the large potential costs of dealing with the impacts of unmitigated climate change are included.

However, even from the most optimistic view, there is a very long way to go before these collective benefits are seen as sufficient to motivate individual companies. Moreover, by itself, a shift to greener products and cleaner production processes and energy technologies will not necessarily resolve the inequalities in world trading patterns. Indeed the opposite might happen. As the early enthusiasts for alternative technology warned, it could be that the technology might be taken over for commercial purposes but the underlying social and political values might be ignored.

To be fair, some companies have responded to the idea of sustainable development in a wider sense by developing new approaches to trading with and investing in developing countries (Schmidheiny 1992). Unfortunately, they may be in a minority. More generally, far from sustainability being a major concern, many environmentalists would argue that in fact the overall trend within the global economy is in the other direction, as a result of the drive towards 'free trade' around the world and the advent of a 'globalised' economy that is dominated by multinational companies. Certainly, so far, environmental concerns have not figured significantly in the World Trade Organization's free trade arrangements. In many environmentalist eyes, there does seem to be a fundamental conflict between free trade and environmental protection – unless environmental controls can be imposed globally.

Even so, as we have seen, there are some signs of change in the way companies are approaching technology since some of them can see benefits from adopting more sustainable approaches. This at least is a start.

Green employment

While companies may gradually adopt a more sustainable approach for commercial reasons, or be pressured to by governments, there are some potential benefits associated with a shift to sustainable technology that may provide an incentive for wider support for change. For example, studies have suggested that at least as many new jobs could be created as would be lost due to the impact of tight emission controls on industry and the phasing out of environmentally undesirable technologies.

Certainly, in recent years one of the main selling points for renewable energy has been its potential employment impacts – replacing jobs in traditional energy employment. For example, the World Wildlife Fund has calculated that the development of policies on renewable energy and energy efficiency could create 1.3 million new jobs in the USA by 2020 (WWF 2002).

Table 14.1 *Direct employment in electricity generation in the USA (jobs per thousand GHh per year)*

Nuclear	100
Geothermal	112
Coal (including mining)	116
Solar thermal	248
Wind	542

Source: Christopher Flavin and Nicholas Lenssen, 'Beyond the petroleum age: designing a solar economy', Worldwatch Paper 100, Worldwatch Institute, Washington, DC, December 1990 (http://www.worldwatch.org)

It is sometimes argued that you can actually get more jobs from renewables and energy conservation technologies than from conventional energy technologies. Certainly the statistics sometimes seem to imply that: see Table 14.1 for a typical set of data.

Care has to be taken when making estimates of the job creation patterns associated with any form of investment. In addition to direct operational employment, on site, there are the jobs associated with manufacturing the hardware, and indirect jobs elsewhere, providing the materials and components. The people in all these jobs will also create further jobs, 'induced' in the economy as a whole, when they spend their earnings. In terms of the conventional economic viewpoint, ultimately, therefore, taking the economy as a whole, the total number of jobs per £ or $ invested will end up being similar, regardless of the initial investment. Put at its most severe, on this view, you will only get more net employment if you either invest more money or pay less wages. However, even on the conventional economic view, you may be able to create more direct jobs *quickly* by investing in labour-intensive technology, and certainly the dynamic pattern, location and type of job creation will differ for renewables and energy efficiency. Ensuring that it matches the existing pattern of employment and skills as well as possible is a matter of careful planning.

There could still be some specific areas where there would be dislocations and a need for retraining. A shift to a sustainable future would clearly involve a need for careful planning to limit disruption. However, the areas of likely growth are those that are currently in decline. For example, aerospace workers could find employment in the wind power field, shipyard workers could work on wave energy technology and construction workers could find employment on tidal energy projects. Certainly, the changeover would involve a lot of new investment and this might be one way in which currently unemployed people could find work. And the changeover might provide more sustainable employment – less threatened by the economic ups and downs resulting from, for example, oil price variations.

Not surprisingly, trade unions have shown interest in such possibilities. Indeed, some trade union groups were among the pioneers of the idea of shifting to 'socially and environmentally appropriate production', as witness the campaign mounted by trade unionists at Lucas Aerospace in the UK in the 1970s. They argued that rather than having to rely on defence-related production, job security could be better ensured by shifting to socially needed products and systems, and they developed their own plan outlining 150 new products that they felt they should work on. These included wind turbines, solar energy devices and fuel cells. Several other trade union groups in the UK subsequently developed similar plans, as did some labour organisations in the USA, where a strong campaign for 'defence conversion' was mounted by workers in some parts of the defence industry (Wainwright and Elliott 1982).

These early initiatives were at a time when the idea of sustainable technology, based on renewable energy and environmental protection, was in its infancy and in most cases company managements repressed or ignored the plans that emerged from their workforces. Indeed some of the activists involved in the UK were dismissed, in effect for challenging the management's right to choose what to produce. But times change and, while acceptance of workers' participation in decision-making generally remains low, some of the technical ideas that emerged from these campaigns are now more respectable – indeed many companies seem keen to be seen to be 'green'.

Trade unions in the UK have continued to press for environmentally sound production, although at times there have been conflicts over the future of nuclear power, with workers in the nuclear industry not surprisingly objecting to the generally anti-nuclear views of environmentalists. Even so, in 1986 the UK Trades Union Congress, which covers all unions, supported a moratorium on further nuclear development and in 1988 backed a nuclear phase-out motion (Elliott 1989).

Subsequently, most UK trade unions became increasingly supportive of renewable energy and clean technology development, as one way to secure safe and worthwhile employment (TUC 1989; NALGO 1990) and, in general, rather than being seen as leading to employment cutbacks, environmental policies are viewed as one way to ensure sustainable employment. Although trade unions have increasingly come to support renewable energy, it has to be said that some are also prone to hedging their bets by also supporting nuclear power, as the UK engineering union Amicus, has done (Jackson 2002).

While it is reasonable to expect overall job gains from new environmental programmes, in the longer term, when and if a sustainable energy and supply system were established, then overall employment might stabilise or even decrease, at least in the primary energy generation sector. It could be that there will be a compensatory shift to more work in the service sector, for example, on repair, renovation and recycling of energy-efficient long-life products and systems, and a shift generally to shorter working hours. However, leaving long-term speculations aside, in the shorter term, making the transition to a sustainable energy system would seem likely to involve a burst of employment. It could be that the 'jobs' issue could provide a way to resolve some of the conflicts outlined in the model described in Chapter 1, and in particular the conflicts between producers and environmentalists.

To summarise, the move to green technology and a sustainable energy future will involve many difficult challenges in terms of overcoming vested interests in the economic status quo. But far from being threatening, developing the necessary technology could have positive economic implications. If there is to be economic growth, this is one area where it could occur.

Some companies have already shown an interest in developing clean technology, and some governments are taking the issue of sustainability seriously. The next chapter widens the focus to look at what changes might be necessary, in terms of the overall global pattern of industrial and technological organisation, if sustainability is to be achieved.

Summary points

- Technical fixes may not be sufficient to achieve full global sustainability.
- There may also be a need for qualitative and quantitative changes in consumption patterns.

- Lifestyle changes and less reliance on economic growth may also be necessary.

- While making the transition would be difficult, a shift to sustainability may lead to improved quality of life and more secure employment.

- Employment gains may provide an incentive for making a shift to sustainable technology.

- Environmental issues are global, and the changes that are needed in order to attain sustainability are likely to have to occur on a global basis.

Further reading

Ted Trainer's book *The Conserver Society: Alternatives for Sustainability* (Zed Press, London, 1995) is a useful corrective for those who feel that technology alone can solve our environmental problems. In similar vein, Ernest Braun's *Futile Progress* (Earthscan, London, 1995) argues that 'progress' as currently defined is an illusion, and that technology should be redirected to meet real needs rather than being driven ever more rapidly to create profits. The critical attack on consumer society has continued, with a recent contribution being Robert Wollard and Aleck Ostry's *Fatal Consumption: Rethinking Sustainable Development* (UBC Press, Vancouver, Canada 2000). However, as the title of this book suggests, the emphasis has begun to move on to trying to define a more positive form of sustainable consumption. A useful contribution to that is provided by Brian Heap and Jennifer Kent's (eds) recent report 'Towards sustainable consumption: a European perspective' (Royal Society, London, 2000), which can be accessed at http://www.royalsoc.ac.uk/policy/sustainreport.htm.

The view from the business world has also undergone some significant changes in recent years. One of the pioneering studies was *Changing Course: Global Business Perspective on Development and the Environment* by Stephen Schmidheiny and the Business Council for Sustainable Development (MIT Press, Cambridge, MA, 1992), which argued that industrialists are already accepting the challenge of developing sustainable approaches, while David Wallace's *Sustainable Industrialisation* (Earthscan, London, 1996) presents a forceful case for relying on international companies to restructure industrialisation on a global basis along more sustainable lines. A more recent review of the debate is provided by Richard Welford and Richard Starkey (eds), *The Earthscan Reader in Business and Sustainable Development* (Earthscan, London, 2001).

The UK government's White Paper on energy, 'Our energy future: creating a low carbon economy' (February 2003) committed the UK to trying to reduce carbon dioxide emissions by 60 per cent by 2050, the first such commitment made by a government.

15 The global perspective

- Global economic development patterns
- The industrialisation process
- Post-industrial society
- The potential for leapfrogging

Environmental sustainability is a global issue: pollution does not respect national boundaries. The industrial system that creates it, and makes use of environmental resources, is increasingly organised on an international scale. This chapter explores the way in which industrial development patterns have influenced energy use around the world since the industrial revolution. It argues that new patterns of technological and industrial development may be needed in both the developed and the developing countries if the environmental limits that face the existing industrial system are to be overcome.

World development

Our analysis so far has suggested that, in very general terms, a sustainable future is technically feasible but socially and politically problematic, in that it might be seen as challenging the industrial and economic status quo. Our focus has mainly been on energy use in the developed countries, which historically have created and continue to create the bulk of the world's pollution. However, the developing world is catching up. Indeed, as Figure 15.1 illustrates, in terms of carbon dioxide production, the developing countries are collectively beginning to catch up with the industrial countries. They could overtake substantially within a decade or so, especially if, as this projection assumes, the industrial countries manage to make some reduction in emissions. Note that, on this

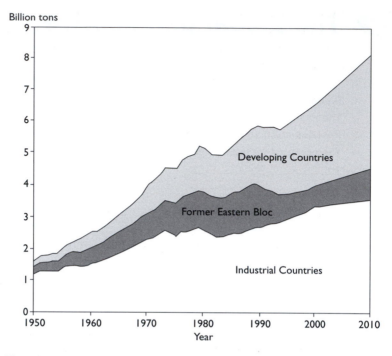

Billion tons

Figure 15.1 *Growth in carbon dioxide emissions from the industrialised and developing countries*
(*Source:* Based on data from the Worldwatch Institute ('Climate of hope', Worldwatch Paper 130, 1996) and, from 1990 onwards, from the DOE/EIA International Energy Outlook, 1999)

projection, emissions from the ex-Soviet countries also begin to increase again. Clearly, there is a need for worldwide remedial action.

Voluntary commitments to making cutbacks may of course emerge, but in general it seems that unless agreements on emissions can be negotiated on an international basis there is little hope for long-term environmental sustainability. However, the developing countries may see little reason why they should not follow the industrial countries' path to affluence. But can and should they?

That opens up the complex issue of global economic development, and in particular, the question not just of whether sustainable growth is possible, but also of the distribution of any growth between the rich and the poor of the world – the North and South.

Recently, there has been much debate (e.g. at the 2002 UN Earth Summit in Johannesburg) on how 'sustainable development' can be achieved worldwide and on defining what that means for economic growth, trade and aid – not least in terms of transfer of technology from North to South. Renewable energy would seem to offer an important option. If advanced industrial countries can, given time, conceivably shift to renewables, then it does not seem in principle unreasonable for developing countries to follow that path, rather than copy the route that the West adopted historically. Renewables are well suited to many Third World countries, but whether it is possible for developing countries to 'leapfrog' over the dirty and inefficient intermediate stage of industrialisation remains to be seen.

A key factor shaping whether energy demand can be met in the future on a worldwide basis in an environmentally sound way will be the scale of population growth and the success of strategies for constraining it. However, that in turn depends significantly on the level of economic development that is possible: wealthier populations tend to have smaller families, assuming they have access to birth control methods. This is not the place to explore the complex issue of population control. However, it does at least seem possible that if new energy technologies like renewables can help developing countries to meet their energy needs without having to import expensive fuel or destroy their own environmental assets and rural economies, then perhaps the development process can create and distribute wealth more widely. Population growth may then be stabilised, this in turn making it possible for renewables to meet energy demands.

So far in practice within the developing world, renewable energy technology has, in many cases, been relegated to a relatively minor role as an 'Intermediate Technology', with many countries seeking, instead, to copy the historical Western model of fossil-fuelled, or in some cases, nuclear development. But given the right sort of technical and financial support, alternative routes may be feasible that are perhaps less economically draining and environmentally damaging.

There is always the temptation to opt for what seems to be the easiest, quickest and cheapest route and some key developing countries have access to substantial reserves of dirty fuel, such as brown coal. But international agreements can be developed to try to ensure that a more environmentally sustainable approach is adopted. This will not be easy but in principle international agreements, if conformed to by all countries, can help to ensure that the extra cost of abiding by agreements does not put participant countries at an economic disadvantage.

The Kyoto climate change accord, negotiated initially in 1997, is one attempt to start this process. However, as the battles over the Kyoto accord have illustrated, not all countries will necessarily want to, or be able to, subscribe to international agreements: this is a matter for careful diplomatic negotiation. For example, in some cases international agreements are phased or regionalised to take account of different levels of economic development. Thus, as an interim measure, developing countries may be allowed to release more emissions than the developed countries, since the latter have not only produced the most pollution so far, but also have the technical and financial means (and, one might add, duty) to bring about improvements. This was what happened with the Kyoto accord: it only applies formally to the industrialised countries; and the newly industrialising countries and the less developed countries are only asked to join in on a voluntary basis. At the same time, the international agreement may propose aid plans, and technological transfer initiatives designed, like the Clean Development Mechanism, to help the less developed countries to develop and use cleaner technologies.

To try to explore the way in which developing countries might move ahead, it may be helpful to look at the energy dimension in the pattern of world industrial development so far.

The post-industrial world

A gradual transfer of emphasis from energy- and material-intensive industries to an economy based on a reduction of such activities is typical of most mature industrial countries. Industrialisation involves a rapid increase in the use of energy in any given country, because it is building up the basic industrial infrastructure. The emphasis is typically on the extractive industries (e.g. coal mining), steel production and the manufacture of basic commodities.

Subsequently, however, as Riccardo Galli has put it in a study of 'Structural and institutional adjustments and the new technological cycle', as industrial countries mature, 'quantity gives way to quality, light industries and services tend to dominate the economy, so that the need for materials and (energy) services declines' (Galli 1992: 781).

Industrialisation thus gives way to the development of a *post-industrial society*, with the emphasis less on primary 'heavy' industries and manufacturing and more on services, with communications technology playing a key role. In particular, large-scale energy-intensive production of materials such as steel decreases, as has happened in the UK, USA and Japan. Of course, initially at least, these dirty primary industries may simply relocate in newly developing countries, typically because labour is cheaper and environmental controls may be less rigorous. However, the development of new technologies may also change the overall pattern of production. For example, new materials have been developed that require less energy to fabricate, and they are being substituted for conventional materials in increasing numbers of products. Thus use of plastics, ceramic, carbon fibres and composite materials replaces the use of metals in an increasing number of applications. As the use of new materials spreads, gradually the overall global level of production of traditional materials decreases, as does the energy required for the production of materials. The process is sometimes called *dematerialisation*. You may remember that this was the title of one of the Shell scenarios discussed in Chapter 9 (see Figure 9.2).

In some countries elements of the 'post-industrial' trend have been underway for some time. The UK was the first to industrialise, followed by Germany and the USA, and you would therefore expect each to reach the post-industrial phase in roughly the same sequence, although there might also be some variations in pace and even some 'leapfrogging'. For example, some latecomers to industrialisation might pass through the various stages more rapidly and perhaps overtake the initial leaders.

Of course, there may be other forces at work, such as shorter term international trade patterns, and longer term structural political and economic changes. But it seems clear that the contemporary process of dematerialisation and the shift to 'greener' technologies and 'cleaner' production systems, which are underway in countries that have reached industrial maturity, are at least in part an attempt to transcend the energy, resource and environmental limits of conventional industrial society. The result of this process could be a transition to a new 'post-industrial' pattern of organisation across the world.

Energy trends

To try to get a feel for the process of industrial change, it is useful to look at the patterns of energy use of some key countries around the world. They are shown in Figure 15.2, in terms of energy intensity, that is, the

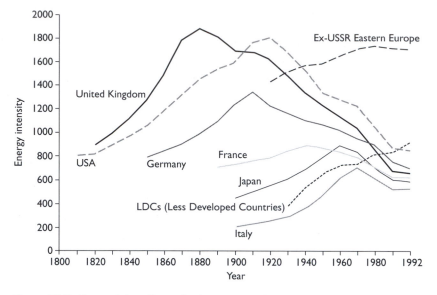

Figure 15.2 *Energy intensity peaks for a range of countries as they have industrialised, in commercial energy/GNP*
(*Source:* U. Farinelli, 'A setting for the diffusion of renewable energies', *Renewable Energy*, 5(1): 77–82. Copyright 1994, reproduced with permission from Elsevier Science)

ratio of energy use to economic activity, the latter being measured in terms of gross national product, or GNP (Farinelli 1994).

Figure 15.2 reveals some interesting trends. It seems that, after industrialisation got underway, the ratio of energy consumption to economic productivity in each country initially increased, then reached a peak, and thereafter decreased, as the country moved towards a post-industrial economy. Moreover, the energy intensity peaks that were reached at some point after each successive country industrialised are all progressively lower. Each country in turn presumably learnt from their predecessors and was able to make use of increasingly more efficient technology: some of the key countries at the forefront of the new innovation cycle were able to use the new technologies to improve their energy efficiency.

The UK's energy intensity ratio (energy consumption/GNP) peaked at around 1880. Forty years or so later, around 1920, the USA reached a lower peak, although Germany had actually achieved a lower peak a little earlier. Forty years or so later, around 1960, Japan reached an even lower peak, followed by Italy – at around half the UK's historical peak.

Kondratiev cycles

The successive energy intensity peaks in Figure 15.2 tie in roughly with the forty to fifty-year cycles of economic peaks and troughs in the world economy first identified by the Soviet economist Kondratiev in the 1920s. Subsequent theorists in the West took up his ideas (especially after the economic crash, on cue, in the 1930s) and some have suggested that the key factor in creating the periodic booms was the development and adoption of a range of key new technologies (Schumpeter 1939). Theories as to precisely how this occurs vary, but, in some versions of what is sometimes known as 'long wave' theory, the technological innovation process is supposed to be stimulated during economic depressions, when companies who had previously simply been able to expand markets for their existing products, find that profits are falling. They therefore invest in new products and a boom results, after which markets begin to saturate once again, so that the cycle repeats (Mensch 1979).

The first Kondratiev cycle is seen as having started with the industrial revolution, chiefly in the UK, based on coal/steam power-related technology. The global economy is currently supposed to be at the end of the fourth Kondratiev phase and entering the fifth, with the main 'new technologies' being information and communication systems, coupled perhaps with biotechnology and renewable energy technology.

Certainly, in the current phase, one of the key factors shaping the development of new technologies would seem to be concerned with the energy resource and environmental limits associated with existing energy technologies. Thus the technologies that emerge as part of the next cycle will, in effect, be those that transcend the limitations faced by the previous set of technologies. To an extent the optimistic views associated with futurologists like Hermann Khan, as discussed in Chapter 14, would seem to hold some force. The limits to growth that were predicted by some environmentalists in the 1970s may in fact be lifted by a shift to a new set of cleaner and more energy-efficient technologies.

However, we have to be careful not to see this as a simple mechanical prediction. Some current versions of long wave theory put the stress not so much on the advent of new technologies themselves as on the social and cultural processes that surround their adoption (Tylecote 1991; Freeman 1992). As our case studies in earlier chapters illustrated, new technologies may not always be fully developed or widely accepted: there is a need for the right sort of social and institutional context, and the process of social and institutional change may have a different timescale from that of technological change. These patterns of social and institutional change may of course reflect economic cycles, but equally there can be a mismatch between the economic cycles and the pattern of social and institutional change.

Not everyone accepts Kondratiev's long wave theory, or its predictions of regular cycles in economic activity led by technological change, and there has been a long-running academic debate on this issue (Rosenberg and Fristak 1983; 1994). However, it does seem clear that there are economic booms and slumps and that innovation does play a role in booms. It also seems possible that energy-related developments are one factor shaping the new technological patterns. The cyclic pattern in energy intensity in Figure 15.2 certainly suggests that.

Of course, the matching of the energy intensity peaks in Figure 15.2 to the industrialisation process is not exact and in some cases is completely absent. For example, Italy, France and Japan were industrialised long before the energy intensity peaks shown in this figure. But in general there has been a similar trend in each country following on from industrialisation. At some point in the subsequent period new technologies are adopted which are more energy efficient.

Note that we are not talking about a reduction in energy use because of economic and industrial decline. Although that may happen in some cases, the graph depicts an improved ratio of energy use to economic output, which implies economic success, but success attained by new means. Thus, typically, there is a shift from energy-intensive 'smoke-stack' industry and mass production, with information-based trading and service sector-based activities becoming increasingly widespread, the latter being, generally, less dependent on energy and material resources.

The most obvious exception is the ex-Soviet bloc, which seems to have just about peaked in energy terms in 1990 at a very high level (of inefficiency); it could be said that its industrialisation has not yet matured. A key issue for the future is whether it can make the necessary technological and economic transition. Carbon emissions dropped rapidly due to the economic decline following the collapse of the USSR in 1990, reaching their lowest point in 1998. Subsequently, as industry and trade recovered, emissions have only increased slightly and, despite continuing economic problems, the overall energy intensity has improved and the prognosis for the future looks reasonable. Certainly, Russia's decision in 2002 to ratify the Kyoto accord is a sign that progress may continue.

The developing countries

What about the countries that have only just started their industrialisation process? The figure suggests that the less developed countries (LDCs) are still on a rising energy intensity curve. The key issue in this context is whether they will be able to learn from the experience of other countries and make the necessary transition. As we have seen most countries so far have been able to improve on their predecessors' performance. It would be tragic if the LDCs could not follow and instead replicated the worst excesses of the first countries to industrialise. It would obviously be preferable if they could 'leapfrog' the conventional industrialisation process, that is, miss out on some of the intermediate stages, and opt straight away for the 'cleaner', more efficient technologies.

What evidence is there so far of any moves in this direction? As was noted in Chapter 8, China is probably the pivotal country, given its size and rapid economic development in recent years. Interestingly, a review by the US Natural Resource Defence Council in 2002 found that China's carbon dioxide emissions fell by at least 6 per cent between 1996 and 1999 while its economy grew by at least 22 per cent. By contrast, the review found that US emissions over the same period grew by approximately 5 per cent. Evidently, then, China has managed to begin to make the transition, and is moving down the energy intensity curve. See Figure 15.3, which shows the reduction in resultant carbon intensity, measured in terms of emissions of carbon dioxide/dollars of economic output. Indeed, as this figure shows, it was able to reduce its carbon intensity faster than the USA. However, that is perhaps not surprising given that the USA has a well-established industrial economy which, as we have seen, went through the energy intensity transition some time ago. The big gains in China seem to have been made mainly by a radical restructuring of the very inefficient industrial system bequeathed from the 1960s and 1970s. Whether this progress can be continued, or is just a one-off gain, remains to be seen. However, the fact that it ratified the Kyoto accord in 2002 suggests that it intends to persevere and continue to reduce its energy and carbon intensity.

Of course, care must be taken not to read too much into the energy/carbon intensity curves as a guide to social and economic development. For example, it could be argued that GNP, or economic output, is not a very good measure of 'progress'. In addition, as noted earlier, part of the explanation for these curves is that, as countries develop economically, they shift to energy- and material-efficient service and manufacturing,

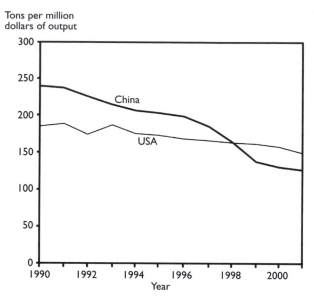

Figure 15.3 *Carbon intensity trends for the USA and China (tonnes of carbon emissions million dollars economic output)*
(*Source:* 'Reading the weathervane', Worldwatch Paper 160. Copyright 2002, http://www.worldwatch.org. (See also the NRDC's analysis at http://www.nrdc.org/globalwarming/achinagg.asp))

and simply export the inefficient 'dirty' industrial activities and technologies to less developed countries, where environmental controls are weaker. The existing industrial countries certainly have more money to invest in cleaner, more efficient, technology, whereas developing countries often have no choice but to use dirty inefficient technologies. It is also clear that companies in the developed countries sometimes 'dump' dirty inefficient technology on developing countries, and set up operations there in order to escape ever tightening environmental controls at home.

There would therefore seem to be a clear need for comprehensive global environmental controls to avoid this risk. In addition it would seem necessary to support 'technology transfer' to the developing countries, so that they too can move up the learning curve – and all nations can adopt clean, energy-efficient technology. Obviously, this will be hard to achieve. Despite regulatory pressures favouring the use of cleaner technologies and the economically defined global trend towards dematerialisation that was mentioned earlier, there will still be areas where older dirtier practices and production priorities continue to be economically attractive, at least in the short term. Economic incentives, such as rewarding companies who adopt cleaner technology with tax concessions, may help to change the balance. However, some analysts fear that, at the very least, even if the developing countries are able and willing to adopt clean technologies, rapid economic growth could still lead to a difficult transition period, during which overall global environmental impacts would continue to increase (Harper 1996). Indeed, some argue that, given that the already industrialised countries have benefited from being able to pollute in the past, it is reasonable and fair to allow the developing economies some room to grow, even if that does lead to pollution. Subsequently however, they must join the industrial countries and converge on the path to lower emissions – the so-called 'contract and converge' approach to sustainable development, as promoted by the Global Commons Institute (Meyer 2000).

Limits to growth

Even assuming a transition to a more energy-efficient, post-industrial, clean technology-based future is successfully made by all developed and developing countries, the convergence zone on the bottom right of Figure 15.2 still implies global economic growth; and overall growth, albeit at lower levels, in total global energy and material use. Dematerialisation and improved resource efficiency can only go so far – energy intensity may reduce, but total energy use, and therefore total greenhouse gas emissions, will still increase as growth continues.

This seems to be the inevitable fate of the so-called 'Kyoto Lite' policy adopted by President Bush in the USA. In 2002, he announced that, rather than backing the Kyoto accord, which called for the USA to cut emissions by 7 per cent, on 1990 levels, by 2008–12, he would introduce a programme designed to reduce greenhouse gas intensity (emissions/GDP) by 18 per cent by 2012, i.e. the reduction in emission of greenhouse gases is linked to the gross national product. By index linking emissions in this way, the USA can continue to have economic growth and also continue to increase emissions. The result could be a net increase in greenhouse gas emissions, over 1990 levels, of perhaps 14 per cent by 2012. Clearly, this was a political compromise, designed to respond to climate change but to avoid any painful cutbacks in economic growth.

If every country adopted this approach, then we could reach the crunch point, where climate change would begin to have serious social and economic effects, sooner rather than later. We are back to the familiar issue of the ultimate limits to growth. Even if the world does make the transition to a post-industrial economy, this new more efficient, but still growing, level of economic activity may not be sustainable on a global basis in environmental and resource terms. So must an attempt be made to reach a steady state economy and a pattern of consumption which is not reliant on growth?

Switching to renewable energy sources would obviously help, but even so there are limits to what these sources could supply in total. The technological limit is defined by the total amount of energy that can be extracted from natural energy flows. As was noted earlier, Gustav Grob has called this the 'natural energy limit'. In terms of long wave theory, that might be seen as the limit for the next Kondratiev cycle, that is, the end of the fifth cycle, although it may take much longer than fifty years to reach it. Equally, even given the adoption of cleaner technologies, if economic growth continues to be the dominant concern, then the limit may be reached earlier.

Obviously, we can only speculate on the longer term possibilities but, to summarise, it does seem that, in the medium term, renewable energy technologies, coupled with the use of Information and Communication Technology and cleaner, greener and more energy-efficient technology generally, can play a key role. They might help us transcend the energy and material limits to further development typical of the technological and economic arrangements of the current type of industrial society. However, there may be limits to subsequent growth, imposed by the limits of the renewable energy resources and possibly by other resource constraints. Technological innovation may help raise the current environmental and resource limits to economic growth, but reliance on renewable energy technology would introduce new limits. In which case, at some point, a new type of technological and social transition would presumably be needed.

What next?

We have, in this chapter, looked at the influence of energy- and resource-related factors on technological and economic development in the past, and it was suggested that environmental constraints, along with economic factors, may be the main influences shaping the pattern of technological development in most sectors of the global economy. However, the precise direction and pace of change is hard to predict.

Economists like Freeman make use of the Kondratiev long wave theory, but have increasingly tended to down play the idea that there are fixed fifty-year time periods. Instead the emphasis is on 'phases' and transitions, and the adoption of new techno-economic paradigms, stimulated by social, economic and environmental pressures, without any specifically determined dates (Freeman 1992).

As has been suggested earlier, it may be that the next phase, the so-called 'fifth Kondratiev', will last longer than fifty years. For example, in purely technological terms, it would take at least that long before renewable energy could make a major contribution to meeting the world's energy needs. Current energy predictions suggest that, if a major commitment was made to developing renewable energy technology, it would take between fifty and one hundred years before renewables could supply half of the world's energy. So it could be some time before the 'natural energy limit' shown on Grob's chart (Figure 3.1) is reached.

Of course, it could be that, well before the natural energy limit is reached, other resource constraints will impose limits on economic activity that cannot be transcended just by 'technical fixes' like the adoption of renewable energy technology. Technological innovation can certainly lead to economic growth but it may also, in the end, be unable to sustain it. In which case there might be a need for possibly painful social and economic choices if not immediately, then further ahead.

For the moment, however, the task would seem to be to start on the long process of developing sustainable energy technology. In the next chapter, to begin to round off our discussion, we will look at some of the more specific practical technological choices in the renewable energy field. This should highlight some of the constraints that lie ahead but also indicate some of the opportunities for positive changes in society as well as in technology.

Summary points

- The Western industrialisation process cannot be repeated worldwide without major environmental consequences.

- Fortunately, technical fixes and new technological development patterns may reduce this problem.

- However, there may still be longer term environmental and resource limits on the extent to which economic growth can continue.

- For the moment we are faced with a range of choices over how to develop sustainable energy technology, as part of the process of moving to a sustainable future.

Further reading

The debate on global futures has a long history – going back at least to Donnella Meadows, Dennis Meadows and Jorgen Randers' seminal *Limits to Growth* (Earth Island, London, 1972). These authors in effect updated their analysis in *Beyond the Limits: Global Collapse or a Sustainable Future?* (Earthscan, London, 1992) and since then there have been a wide range of studies of the future. One interesting text is Allen Hammond's *Which World? Scenarios for the 21st Century* (Earthscan, London, 1998).

The academic debate over Kondratiev long wave theory is complex, but if you wish to explore some of its implications see Andrew Tylecote's *The Long Wave in the World Economy* (Routledge, London, 1991). Perhaps more accessible is Chris Freeman's *The Economics of Hope* (Pinter, London, 1992), which makes use of some of the ideas from long wave theory, and presents an optimistic prognosis.

Much of the debate over global futures is nowadays couched in terms of policies and options for sustainable development, and certainly there is a vast and growing literature on this topic. For a useful introductory review of some of the key issues from a critical perspective see Sharon Beder's *The Nature of Sustainable Development* (Scribe, Melbourne, 1993). More recent overviews have been provided by Michael Carley and Phillipe Spapens in *Sharing the World: Sustainable Living and Global Equity in the 21st Century* (Earthscan, London, 1998), Jennifer Elliott in *An Introduction to Sustainable Development* (Routledge, London, 1999) and by Paul Ekins in *Economic Growth and Environmental Sustainability* (Routledge, London, 1999). For the sake of balance you might also want to look at a book which adopts a more sceptical or contrarian viewpoint on sustainable development, for example, Julian Morris' *Sustainable Development: Promoting Progress or Perpetuating Poverty* (Profile Books, London, 2002).

Finally, for more on the 'contract and converge' concept of sustainable development see the Global Commons Institute's web-site: http://www.gci.org.uk/.

16 The sustainable future

- Alternative paths to a sustainable future
- Choosing technology to fit
- Structuring the social control process

Sustainable development is likely to require more than just technical fixes, but there is no one blueprint for a sustainable future. Instead there is a range of social and organisational as well as purely technological choices. This chapter argues that the process of choice must involve a social process of negotiation over the social and environmental ends, as well as the technological means.

Social choices

The general conclusion that has been reached in the analysis in this book so far has been well expressed by Australian environmentalist Sharon Beder:

> So long as sustainable development is restricted to minimal low-cost adjustments that do not require value changes, institutional changes or any sort of radical cultural adjustment, the environment will continue to be degraded.
>
> (Beder 1994: 18)

We have looked at a variety of technical fixes and at some more radical technological developments. We have also discussed the view that there must also be significant social changes, or at least lifestyle changes. Patterns of consumption may have to change with at the very least more emphasis on conservation, recycling and less emphasis on consumerism.

In the view of some greens, more dramatic changes might also be necessary – for example, social, economic and technological decentralisation. However, equally, it is possible to imagine a basically unchanged society using technical fixes to try to avoid such changes.

The technical fix approach has its attractions, in that it may help to avoid fundamental changes. After all, while some people see radical lifestyle changes as desirable ethically and politically, others would see them as unfortunate impositions. But change of some sort seems inevitable. In the belief that it is vital to try to steer a conscious path towards an environmentally sustainable and socially equitable future, rather than having changes forced on us, it seems wise to try to map out what sort of society might be aimed for.

Utopians through the centuries have tried to do this, by producing blueprints for ideal societies. They can be very inspiring. But they tend to ignore the thorny transitional problem of how to reach the ideal state. In reality what seems to be needed is a process of social negotiation of both means and ends. There is no one obvious way forward: rather there is a host of often complex tactical and strategic issues to resolve and a host of possible ways forward.

Alternative trajectories

The alternative society dreamt of by the counter-culture in the early 1970s had, as its technological base, a set of alternative technologies – the classic windmill, solar collector and biogas generator used in a small self-sufficient rural community. This dream of rural retreat lives on, but has it much relevance to the bulk of the world's population? Some greens think so, for no other reason than they believe that the mainstream society is likely to collapse and so the best place to be is 'not underneath'.

But even assuming mainstream society continues, some sort of social and technical decentralisation would seem to be a valid goal. Large-scale centrally integrated systems, once seen as modern and efficient, are often now seen as being hopelessly inflexible, prone to monopolistic or bureaucratic control, and environmentally undesirable.

Does that mean that small is always beautiful? Certainly it can be. But we must be wary of simple technological determinism: technology does not define society. After all a decentralised society existed in the Middle Ages, based on small-scale wind, water, and biomass, but few people would suggest that feudalism was a progressive state, whatever the technology. Neither does society totally define technology: there are a host of possible ways forward technologically – a whole series of technological trajectories based on different mixes, types and patterns of technological activity. Each may involve a slightly different balance between social and environmental concerns, reflecting different political and cultural attitudes. That is why it is so difficult to prescribe blueprints.

Maintaining diversity is a good ecological principle. There is room for a variety of approaches to sustainability, tailored to specific social, cultural and environmental contexts. However, there are also overall limits to the range of social and technological choices – primarily environmental limits. We have explored some of them in previous chapters in technical terms, like the natural energy limit to total amount of renewable energy available; and we have outlined some general criteria for sustainable technology based on recognition of this sort of limit. But within this 'sustainable development' envelope, there is a range of social and technological possibilities.

A key task for the future is to try to reach agreements on the right mixes in each situation, locally, nationally and globally, while recognising that each of these levels of concern interact to some degree. This will involve some trading off between short- and long-term concerns, tactical and strategic considerations, and local and global perspectives.

Small is beautiful?

In order to make these rather abstract generalisations more concrete, let us look at the issue of technological scale, since it provides an example of the problem of balancing out some of these conflicting local and global, tactical and strategic concerns. The attractions of small-scale technologies, as initially promoted by the alternative technology movement in the 1970s, seem clear enough. Small-scale systems avoid losses in distribution from centralised power plants, and they seem likely to have lower environmental impact than large systems. They can also, in principle, be built and run by individuals or small groups, and they can therefore perhaps foster local self-determination and democratic control.

However, it is not clear if these potential benefits always materialise in practice. For example, local control may not always lead to optimal environmental choices in wider national or global terms. Remember the debate over wind farms in the UK. NIMBY (not in my back yard) reactions may dominate to resist some potentially beneficial technologies, while other technologies of arguably less merit may be welcomed for purely selfish local reasons. It is not always the case that local residents take care of their environment.

Secondly, small-scale technology is not always environmentally appropriate. The basic physics of wind turbines implies that small-diameter machines are much less effective that larger machines: as you may recall, there is a square law for power output, so doubling the blade diameter quadruples the power output. Small machines can perhaps be more easily located near human settlements without having an adverse impact, but using larger machines, especially if they are grouped in clusters, can generate sufficient power to make it possible to site them remotely and send the power back to users on power grids. Remote location makes it possible to get access to high wind speed upland sites and, as you may remember, the power in the wind is the cube of the wind speed. So bigger machines in windy sites are very much more effective than the same total capacity of smaller machines at less windy locations.

This is reflected directly in the cost of generating power from small devices. A typical 1 kW domestic-scale wind turbine costs around £3,000. Typical commercial-scale wind farm machines of say 1 MW cost around £600 per installed kW, that is, five times less per kW and even less per kWh of electricity produced. The extra cost reflects the fact that small machines use more materials per kW installed and per kWh produced, than big machines – more steel for the towers, copper for the generator and so on. A study of the 'energy balance' (i.e. input energy versus output energy) of a range of energy generation technologies compared the amount of energy needed to manufacture them (and their constituent materials) with the amount of energy they produced in operation. It indicated that small 'stand alone' wind turbines, using battery storage, came out as the worst option of all (Mortimer 1991). So people who choose to run small wind turbines to try to attain some degree of self-sufficiency are in fact not only paying a cash premium, they are also imposing an extra environmental resource cost on the planet.

There is a further problem in terms of storage. The individual 'off-grid' wind turbine, run independently to supply power to a remote dwelling, will need some form of power storage, since the wind does not blow all the time. Batteries are widely used. But they contain lead – do we really want to bequeath our children yet more lead to deal with? By contrast, with a national network of grid-linked wind farms, power can be shunted around the country, with each wind farm supplying power to the grid when it can. The grid can thus act as a form of buffer or store to compensate for local variations in wind power availability. That should more than offset any power losses due to transmission along the grid.

There are situations where local independent 'stand alone' generation is reasonable and indeed sensible – in 'off-grid' locations where the alternative is to use diesel or petrol to fuel generators. There is a lot of interest in this level of activity in remote areas. However, there is also increasing enthusiasm for independent power amongst people who want to be self-sufficient, even though they might be able to use grid power. In the USA in particular there has been something of a rebirth of the 1970s alternative technology movement, with the emphasis being on 'getting off the grid' via self-help initiatives (Jeffrey 1995). This option is somewhat easier these days due to technological developments; for example, some 'home power' enthusiasts are using photovoltaic cells and other high-tech devices (Allen and Todd 1995).

The more traditional approach, as pioneered by the old alternative technology movement, involved putting more emphasis on 'low technology', for example, making use of scrap materials to construct simple wind devices on a local do-it-yourself (DIY) basis, and thus avoiding the use of scarce resources. Certainly, one of the still valid principles of the old alternative technology movement is that it is environmentally helpful to try to meet local needs from local resources. However, there are limits. It is probably much easier in terms of, say, biofuels like wood than for more complex engineering-based generating technologies like wind turbines, unless you are prepared to use relatively simple perhaps less efficient homemade devices. There is clearly a role for local-level DIY initiatives, but overall it seems likely that, in terms of small-scale local usage of renewables, the future will lie mostly with professionally designed and built high-performance machines and systems.

Off the grid?

Assuming that most of the wind machines used will be professionally designed, this still leaves the question of whether they can and should be run independently, that is off the grid. The analysis of scale factors above suggests that totally 'off-grid' generation, for example, using small wind turbines, may be environmentally suboptimal as well as expensive, with the implication being that it would be better to share common services, for example, via wind turbines linked to the grid, so as to trade between areas and even out local variations in wind availability.

This conclusion may be changed to some extent by technological breakthroughs. In terms of cost, mass production of wind turbines is likely to reduce costs and, traditionally, cost reductions are most when you are dealing with larger numbers of identical mass-produced products. So smaller machines may get cheaper faster than big turbines. Improved designs may also use less material resources. The development of new energy storage techniques and a move towards the use of hydrogen as a new fuel might also make reliance on local energy sources more credible.

Organisational changes might also improve the situation for smaller scale technology. In the UK, for example, following the liberalisation of the energy distribution network, it is now possible for consumers to contract with any supplier, using the grid as a 'common carrier'. Previously, the only option was to buy power from the regional electricity company. Under the new arrangements, as we noted earlier, the supply companies contract to match the power they sell consumers with power bought in from renewable generation companies on a kWh per kWh basis. Green power retail schemes of various types have emerged around the world, as markets have been liberalised. See Box 16.1.

Box 16.1

Green power retailing

Most industrialised countries now have some form of green power retailing – with electricity companies offering to match the power sold to consumers with power bought in from renewable energy generators. Some companies charge a premium price for this service – up to 15 per cent extra in some cases – although others have managed to avoid a surcharge. In some cases in the UK, the company offers to use some or all of the surcharge to support a *fund* for the development of specific renewable energy projects, often smaller scale local projects, so that consumers can identify directly with them.

The scale of consumer uptake of these various offerings has varied around the world. Initially, there was a lot of enthusiasm for the idea of what are termed in the USA 'green tariff' schemes, with initially, in the *USA*, over

100,000 consumers signing up after the launch in 1998. However, the Californian energy crisis in 2000–1 in effect killed off most of the schemes there, although progress continues to be made in other states. Australia also launched green tariff schemes in 1998, and also saw quite large initial uptake. In *Europe* the lift-off was initially slower but it grew quite rapidly. By 2002, around 324,000 consumers had signed in Germany and over 1.3 million in the Netherlands, around 14.5 per cent of the population.

The UK was the laggard with only 60,000 consumers participating by 2002, with one of the main problems being the lack of capacity available for the green power market – so that most energy suppliers have tended not to advertise their schemes very strongly for fear of being overwhelmed with demands they could not meet. Moreover, under the rules of the Renewables Obligation, only power from projects that is additional to that being claimed against the supply companies' Obligation, can be used for the domestic green power market, since otherwise consumers who volunteered to buy into a green power scheme might be double charged – first by the extra cost of meeting the RO, passed on to them by the supply companies, and again by any premium charged for the green power scheme.

Some UK energy suppliers have tried to work round the problems outlined above by moving over to 'fund' schemes, or hybrid fund–supply schemes, but so far the level of uptake remains low, as does the amount of new capacity that has been installed on this basis, that is, outside the RO. Some people have argued that, in any case, it is more equitable for *all* consumers to contribute to developing renewables, as with the RO, rather than leaving it to a few altruistic individuals. On the other hand, there are clearly consumers who want to make a personal commitment and the 'voluntary' schemes provide a way. Certainly, that is the message from much of the rest of Europe.

Be that as it may, most of the green power that has been available so far in the UK has gone not to the 'voluntary' domestic green power market, but to meet the demand from the business sector created by the Climate Change Levy, since power from renewables is exempted from the levy. In its first year (2001–2) that led to around 6.8 TWh in exemptions. In addition to companies, subscribers so far have also included many local authorities, universities and community organisations. Some of these organisations have a direct interest in the development of renewables – for example the Co-op Bank was one of the first to subscribe and is involved with a range of ethical investment activities. Similarly, the Body Shop, which is well known for its ethical stance, has been subscribing to a green power scheme for some while, and in 2002 it linked up with *Ecotricity*, one of the green power pioneers, to offer its customers access to a voluntary green power scheme.

For regular updates on the green power market around the world, see the greenprices web-site: http://www.greenprices.com

In principle anyone can feed power into this green power market, even homeowners who have excess green power to sell, assuming they meet the technical standards. In practice, it has been the existing power and distribution companies that have made most of the running in the UK and Europe, but some smaller independent companies have also emerged. For example, in the UK the Renewable Energy Company was set up in the Stroud area in the West of England, initially as a local retailer, using power from local generators. Subsequently, under the name Ecotricity, it has expanded to offer green power nationally, and several other independent green power retail organisations have emerged in the UK.

Clearly, the success of such schemes will depend on market conditions – and the availability of green energy supplies at reasonable prices. But in principle it should be possible for excess power from a wide range of small local sources to be used for the 'voluntary' green power market, as well, of course, as feeding into conventional power market.

This type of approach would represent a compromise between total self-sufficiency and national-level integration via the grid. In principle you can get the best of both, with a distributed network of small projects being linked up by the grid, feeding power to local consumers and also to consumers around the country. On the consumers' side, individual households can contract with suppliers of their choice. On the power generation side, rather than having totally independent machines, many individual wind turbines in a region or country, or even more widely, can be linked up to export any excess power to the national grid, so that there can be an integrated yet also decentralised network of power generation – local generation contributing to collective national power availability and collective provisions being available for local use. As was suggested above, local generators could well proliferate in this situation. Indeed, as in Denmark, some local consumers might decide to set up co-ops themselves to supply their own needs and sell any excess power via the grid, taking in power when the level of local generation was not sufficient.

Local generation

The concept of local generation integrated into a national system is an attractive one, especially since it does not need to stop just with wind turbines. Local excess power from local biomass-fuelled plants, or from small hydroelectric turbines in rivers and streams, or from photovoltaic solar cell arrays on local dwellings and commercial premises, could also be added in. Micro hydro is of course a site-specific option, available only in some locations, but energy crops and photovoltaics seem likely to lend themselves to widespread use on the basis of local-level decentralised operation, integrated with the grid.

Local generation could be carried on a small- to medium-scale commercial basis, or in, say, the case of PV solar, just at the private house level, making use of the 'net metering' – with the grid company just charging for any net flow of power to the consumer, or paying the consumer if the net flow of power is the other way. This idea has already caught on in some areas in the USA, especially in California where PV solar is being adopted by consumers who, following the power crisis there, no longer felt able to rely just on getting power from the grid. They wanted to generate some of their own. See Box 16.2.

Box 16.2

US consumers opt for PV solar

As noted in Chapter 8, in 2000–1, following the deregulation of the electricity market in California, the power companies were unable to meet demand reliably. With regular power blackouts and electricity bills hiked by 40 per cent or more, some consumers in California turned to PV solar power, even though it was still expensive. To some extent it was a protest action. One consumer told the local Orange County Register 'I'm outraged that large corporate powers can band together and apparently charge whatever they want for power. We're all hostage now to a consortium of electric-power suppliers.'

But it also began to look like a realistic and increasingly popular energy option. The California Energy Commission, which offered 50 per cent rebates on the cost of installing PV systems, took 450 applications from solar buyers in the first two months of 2001, nearly the same amount as received in the three previous years combined, and overall it saw interest in solar systems rise by 500 per cent in the first two years after the energy crisis.

The provision of state-wide and local support schemes clearly helped. The Los Angeles Department of Water and Power (LADWP) introduced a quite generous Solar Rooftop Incentive Program, which subsidised the purchase and installation of a grid-connected PV for a customer's home or business. The incentive amounted to $3 per watt for systems manufactured outside the city and $5 per watt for systems manufactured inside LA city lines. LADWP paid up to $50,000 for each residential system and up to $1 million for a commercial installation.

continued

However, there were some conditions. The household systems could supply no more than 100 per cent of the customer's own power needs, averaged out. Customers could have net metering – a meter that runs backwards – but were not allowed to actually make a profit by selling power, over and above what they consumed, to LADWP – or to anyone else. And, the incentive did not cover the purchase of energy storage systems – systems had to be grid-connected only. From LADWP's point of view, they got more solar generating capacity installed, using consumers' rooftops, partly funded by the consumer, and they could resell any excess solar electricity to other customers on the grid. But clearly not all consumers would be happy with these restrictions – some might want to be able to generate excess power for others to buy, but of course then they would not get the installation subsidy and would, in effect, have to become independent power producers.

For an interesting and often irreverent US analysis of the local generation issues see *Home Power* magazine (http://www.homepower.com), which, in the midst of the California blackouts, had a quip from an anonymous off-grid renewable energy user: 'Blackout? What blackout?' Also see David Morris, 'Seeing the light: regaining control of our electricity system', Institute for Local Self Reliance, http://www.ilsr.org.

The switch to PV is not an isolated trend. Increasingly we are seeing a move to what is being called 'micropower', with micro generators of various sorts, including fuel cells and micro-CHP generators, being installed in many houses and other buildings, so that in effect they become power stations. If this trend continues, we may see the development of a new pattern of energy generation and use. Buildings might generate the bulk of the energy they needed from local renewable sources, including from PV cells on the roof, and import power from the electricity and/or gas grid when there was a shortfall. It might be argued that such a system would not be very cost effective, given the small amounts of power coming in at each point in the network and the losses in transmission along the grid. As one critic put it, you can end up just warming the wires. But, under most conditions, the bulk of the power would actually be used locally, and so this problem is lessened. Indeed it has been argued that meeting local needs by local generation from renewables should actually be given credit since it avoids the power losses associated with distributing electricity from centralised power plants to remote users.

The concept is known as 'embedded generation': local generators are located at the far end of the grid where grids are weak and where the power they produce is more valuable than that which has to be sent long distances from centralised power plants. Obviously, the benefits of using 'locally embedded' generation will only accrue when local sources are producing a part of the net national energy requirements; that is, mainly just what is needed locally. Once power has to be shifted around the country, either into or out of localities, this advantage is lost. But it seems to be catching on.

Some local energy needs could also perhaps be met without requiring high-quality and totally 'firm' grid supplies. For example, it seems foolish to use expensively generated frequency- and voltage-stabilised grid electricity just for heating. In reality, only a relatively few domestic and commercial devices actually need 50 cycle ac mains (most electronic devices convert it internally to low voltage dc) and some devices like freezers can cope with periods of disconnection. It could be that separate local grids will be developed for some uses, perhaps using lower voltage direct current generated from local renewables, topped up when necessary with power from the national grid.

More radically still, as mentioned earlier, there seems likely to be a shift to using *hydrogen* gas as a new energy transmission and storage medium, with the hydrogen being generated via electrolysis of water using renewable sources of electricity to produce the power. That would not only resolve the intermittency problem that will face renewables when they expand to be used on a large scale, but it would provide a new clean fuel that could be used direct for heating, to raise steam for electricity generation or to power a fuel cell to produce electricity directly.

Local co-ops or even individuals could make use of this option – converting any excess power from their own micropower system to hydrogen, which they could either store or trade with other users, via the gas grid. As noted in Chapter 8, hydrogen can be admixed to conventional natural gas (the mixture is known as hydrane) and it may be that this will be done on a wide scale, as methane becomes more expensive, on the way to a new gas distribution system based on hydrogen. See Box 16.3.

As can be seen there is a wide range of possibilities for the future of energy use and a wide range of electricity-generating renewable energy technologies, some of which could be used effectively at local level on a relatively small scale in rural areas, and some of which, like PV solar, could also be relevant to cities. In addition, it is worth remembering that electricity is only needed for some applications: the bulk of local energy needs are for heating, and there is a range of heat production options at local level, including most obviously solar heat collectors, but also a range of biofuel options. For example, it could be that methane gas from biogas generators using local wastes or local energy crops, is fed into the

Box 16.3

Hydrogen co-op

In a paper to a recent energy conference, Japanese researcher Takeshi Kaneda outlined a 'Proposal for a renewable-based hydrogen energy system'. It involved a feasibility study for the idea of using excess power from a wind turbine and PV solar units in a small rural co-op in Japan, to generate back-up hydrogen supplies for use directly and/or for trading via the gas grid. He envisaged a user co-operative of fifty households, each consuming 400 kWh per day, with 3-kW rooftop solar panels on each rooftop and with one 200-kW wind power generator. Any excess electrical power over and above the co-op's needs could either be sold or be used to produce hydrogen gas. This would be stored for later use in a fuel cell to generate electricity when needed, or be traded with others in the region along the gas grid. The choice of whether to sell hydrogen or electricity would depend on what was most in demand and, consequently, the price that could be obtained.

Kaneda found that the system could in theory meet most of the community's energy needs and more, most of the time, although there might be an occasional need to import power, but that the economic viability of the system depended very much on the prices that could be obtained for the traded power outputs. Even so, given that the PV and wind system would generate on average about four times more renewable electricity than the co-op needed, about 75 per cent could be converted to hydrogen and exported as either hydrogen or electricity at peak demand time when prices were high. On the basis of then current utilities prices he calculated that, very roughly, the original cost outlay for the system could be paid back in around twenty years from electricity sales, but less if the hydrogen could also be sold, and even less if a carbon credit system of some sort was in operation.

An even more ambitious idea, linking the local to the regional level, has been developed by a group of Japanese and Russian engineers – a hydrogen pipeline, based initially on a conventional gas pipeline, stretching across the north-west Pacific periphery, linking islands and promontories in a chain from Japan through to Alaska, and fed with hydrogen produced along the way from local wind turbine projects and other renewable inputs.

See K. O'Hashi *et al.*, 'Wind power – hydrogen hybrid systems and natural gas pipeline system in Northeast Asia', NATTA Report, May 2001.

Takeshi Kaneda's 'Proposal for a renewable-based hydrogen energy system' was in a paper presented to the World Energy Council e-conference 'Energy Resources' in 2001, but it is no longer available on-line. For more details see *RENEW*, 135, January–February 2002.

gas mains, and used for domestic heating, or for micro-CHP generation. Local wastes and/or energy crops could also be converted into liquid or gaseous biofuels via local pyrolysis plants and these fuels could be used to meet local heat needs, as well as for powering some vehicles. Moreover, this does not have to be just a rural option, given that cities and urban areas generate so much waste.

Of course, as has been argued above, there are trade-offs to make in terms of economies of scale. There are some renewable energy technologies that can only be used on the relatively large scale – tidal barrages being an obvious example. The barrage at one time proposed on the Severn estuary in the UK would have had an 8.6 GW capacity. Offshore (i.e. deep sea) wave energy generation would be on a similarly large scale. Obviously, there are also possibilities for smaller barrages and smaller onshore or inshore wave energy devices, but it does seem that any viable energy system using renewable energy would have to have a mix of large-scale and small-scale renewable energy technologies. Very roughly speaking, and depending very much on the pattern of society being considered, it might be possible to generate about as much from small-scale local technologies as from large centralised systems. But the precise balance is a matter for debate and negotiation – on the basis of technical, social, economic and environmental concerns.

Energy crops

The need for social negotiation in relation to technical, economic and environmental factors can perhaps be made clearer by looking at a further example of a new form of renewable energy – energy crops. As we noted earlier, the use of biomass as a fuel looks very promising around the world. The specific options vary from region to region: sugar cane has been used to produce alcohol in Brazil for use in cars, while in Europe oil seed rape has been used to produce so-called 'green diesel'. As can be seen, providing transport fuel has been a high priority, but energy crops can also be used as a fuel for direct heat production (e.g. straw is used in many Scandinavian countries to power district heating networks) or for use in electric power generation (e.g. via multifuel boilers or gas turbines). In the UK there is much interest in the regular coppicing of quick-growing willow and poplar to provide wood chips for use in electricity generation via gas turbines (Macpherson 1995).

The use of crops for fuel raises a number of environmental issues. We mentioned some of them in passing in Chapter 7. Perhaps the key

strategic question is whether we should really be using agricultural land
for energy production rather than food production. Part of the reason this
has happened in Europe is that, due to the use of high-productivity
intensive farming methods, there is a surplus in food production. As part
of the EU's Common Agricultural Policy, some land has been 'set aside'
to protect food markets. Using set-aside land for energy crops seems a
better option than leaving it fallow, and it seems better to pay subsidies
to farmers to grow something rather than for doing nothing. Energy crops
are also seen as renewable energy sources – assuming that the rate of
replanting equals the rate at which the crops are burnt, then there is no net
increase in carbon dioxide production, since plants take in carbon dioxide
while growing. However, not everyone is happy with the Common
Agricultural Policy. At the most general level, there are those who feel
that it is obscene to 'set aside' agricultural land to protect profits when so
many in the world are starving, although this of course begs the question
of distribution.

Leaving that issue aside, some environmentalists argue that it would be
better to switch to less intense 'organic' forms of farming, which would
use the set-aside land in a more environmentally sound way (Safe
Alliance 1992). Not only would soil quality and the produce be improved,
but more habitats would be available for wildlife and there might also be
a net energy gain, since the energy used in intensive farming may be more
than would be generated by using the set-aside land for energy crops. The
objectors also fear that energy crops will become big business, so that
there will be giant energy crop plantations dominating the landscape.
They are also concerned about the impact of pesticides and fertilizers –
although the proponents of energy crops argue that the use of these will
be less than with conventional agriculture.

Finally, there are concerns about possible toxic emissions from
combustion. As noted earlier, Greenpeace has mounted a 'Ban the Burn'
campaign, focused on the use of domestic and commercial rubbish as a
fuel for power stations, but environmental opposition could also be
extended to the combustion of energy crops, especially given that some
plants are designed for firing with waste if there is a shortage of wood
fuel. Indeed one such plant, proposed for Wiltshire in the UK's south-
west, has already been successfully opposed by local environmentalists on
these grounds.

For our purposes, what is interesting about this debate is the balancing of
interests amongst the three main groups outlined in our model of

interactions in Chapter 1 – the owners/investors, producers and consumers. Farmers may be keen on energy crops as a source of revenue and employment; investors may see it as potentially profitable; and some consumers may be attracted to the idea of 'green fuel' for cars or 'green electricity' from plants. However, although some environmentalists support energy crops as a useful new renewable source, others are concerned about emissions and local impacts including the land use implications, with potential impacts on water resources being seen as a problem in some areas.

These are not trivial issues: energy crops could become the largest single renewable energy resource. For example, as noted earlier, a study of the UK resource by ETSU suggested that, if fully developed, energy crops like short-rotation coppicing of willow and poplar could eventually provide 150 TWh p.a., or nearly half of the UK's electricity requirements (Department of Trade and Industry 1994b).

This estimate may be optimistic. Certainly, so far progress in the UK has been very slow, and there are many hurdles ahead, even in terms of helping to reach the target of obtaining 10 per cent of UK electricity from renewables by 2010. Whereas, as we saw earlier, the debate on wind farms has now been going some while, the debate on energy crops is really only just beginning. Assuming that some use of energy crops is accepted, then the questions are – how much, in what way, where and by whom. There would seem to be a need to negotiate sensible trade-offs and compromises between the various technical and environmental factors. Local impacts will have to be carefully thought through and discussed, leading to modified planning rules to avoid poor siting and overlarge projects, and to improved emission controls.

Beyond that there would seem to be a need to discuss the overall contribution of energy crops to world energy production. Some energy scenarios, like the global renewable energy scenario produced by Johansson et al. for the 1992 Earth Summit, rely on them heavily; others, like the fossil-free energy scenario produced for Greenpeace, are more wary, with the potentially negative impact on biodiversity being seen as a key issue (Johansson et al. 1993; Greenpeace 1993).

Structuring the debate

Local and global trade-offs between technical, social and environmental concerns of the type discussed above are often at present made primarily in economic terms, on the basis of cost. Attempts have been made to add environmental costs into the equation more effectively, for example by the use of cost–benefit analysis, with financial values assigned to each element. However, this is difficult. How can you value a landscape or a tree? 'Contingent valuation' techniques, in which people are asked to express their individual estimates of comparative value (e.g. how much they would be willing to pay to retain some amenity) are inevitably limited, although they do at least begin to open up some form of social negotiation process.

The economic and industrial forces that shape decision-making about technology are very powerful and it is hard to see how they can be easily counterposed unless there are equally powerful influences at work. In principle, environmental planning legislation and regulatory agencies, such as the Environmental Protection Agency in the USA and the UK's Town and Country Planning system, provide a means for allowing divergent views some hearing, for example, via local public planning inquiries or public hearings on major projects. However, these usually operate within the context of already established higher level technology policies. Thus public inquiries over the desirability of a motorway or a power plant will not normally be allowed to discuss wider issues of transport policy or energy policy, only whether the specific project is acceptable. Even so, environmental concerns are gradually beginning to influence the way governments define their technological and industrial policies, and companies too are beginning to take these issues on board, if for no other reason than to try to anticipate impending government regulations and legislation.

The purpose of this book is not to explore the details of such developments or delve into the rapidly expanding field of environmental economics: this would require a whole book on its own. However, it does seem that criteria for sustainable technology of the sort outlined in Chapter 3 are beginning to be put on the corporate and governmental agenda.

In part this has been possible because, as has been noted, rather than imposing costs on the various human actors in the game, a switch to sustainable approaches may in fact also offer opportunities – for example, opening up new markets for green products, providing employment options for those made redundant from dying traditional industries, and generally creating wealth. Thus, Sir Crispin Tickell, Convenor of the UK Government's Panel on Sustainable Development, has argued that 'People have become rich making a mess over the last two hundred years. In my judgement they could become even richer still clearing it up over the next two hundred' (Tickell 1994: 33).

Not everyone accepts this optimistic interpretation. It is clear that adopting a sustainable approach to technology, and to economic development generally, could create many new jobs and underpin a more stable world economy. However, as we have seen, overall there may be a need for each of the human groups to accept some burden in order to achieve environmental sustainability.

Conclusion

Technology offers both threats and promises. It can ruin the planet, but possibly it can also save it. However, technology is really just a dead assembly of things organised and used, as a tool, for some human purpose. There is certainly a need for careful and concerted assessment in that some technologies may be considered to be socially and environmentally inappropriate, but only the most extreme critic would say that we should do away with all technology. The real need is to be selective and to try to develop a more balanced and sustainable pattern of living, within the constraints imposed by the ecosystem.

The balance is not going to be easy to achieve. Even if people are so minded, it is difficult to know precisely how to defend the environment. After all, although it is interconnected, the natural world is not just one integrated thing. There are thousands of species that may need protecting, thousands of ecological problems that need attention. How can environmentalists choose which to focus on? Must they play god by selecting to defend one species as against another? Is there not a risk that human intervention may become partial, biased and counterproductive?

Some hard-line environmentalists have come out against major renewable energy projects like tidal barrages, on the argument that these would seriously disturb wildlife. The debate is a complex one: while some species might suffer a degraded habitat, others might actually benefit from the local ecological changes. The balance is hard to judge. It could be that other technologies will be a better bet, but there is a need for a full analysis of the options and their impacts. In this situation, the adoption of an overly doctrinaire approach to defending specific species and habitats might be unhelpful.

It must also be the case that some environmental problems are likely to be more serious than others in the long run. Given that there are limited resources of money and expertise, there has to be some degree of focus. But how can assessments be made of which problems are going to be the key ones? The standard argument is that the focus should primarily be on any environmental impacts that seem likely to be irreversible. However, this may not always be easy to identify. What may seem a trivial issue may turn out to be central. In this situation much reliance must inevitably be placed on what scientific research can impart. Unfortunately, however, science can only go so far. In the end it will inevitably come down to subjective human judgements.

Overall, what seems to be needed is a comprehensive approach to environmental protection that seeks to balance all interests and maintain overall biodiversity – and this is not easy. In previous chapters some of the ultimate environmental and resource limits were surveyed and an attempt was made to develop some general strategic principles and criteria for sustainable technology. But this only provides an outline guide for specific choices. A process of social negotiation is needed to make such choices with the natural environment having, as it were, a proxy vote.

Some might say that in a world driven by private and even collective greed it will be difficult to bring about positive improvements. At present, technological and industrial development is being driven at ever-increasing rates by the competitive world economy, fuelled by the profit motive coupled with people's expectations for more and more consumer goods and services. Some people would say that the technology or the rate of change is the central problem, while others would argue, more fundamentally, that it is the underlying social motivations, and the economic and industrial structures that have been created as a result, that are the basic cause (Braun 1995).

There is no way that any objective conclusions on such issues can be reached in this book: it is a matter of viewpoint. Clearly, technology can reflect both the best and the worst human motivations. That is not to suggest that technology is a fixed set of items which can either be used or abused. The way technology is developed and deployed, and the choice of technology is a social and political process, reflecting the conflicting interests and power of rival groups in society. This has always been the case. What has changed in recent years is that, to return to the model of interaction outlined in Chapter 1, a new factor has become crucial – the natural environment. If humans are to live successfully on this planet it becomes increasingly vital that their technological choices take account of more than just the partisan interests of the conflicting human groups or the interests of power elites, and include the fundamental environmental limits and constraints. In practice what seems inevitable is that a process of prioritisation of the issues and options will have to emerge. What is not clear is the political issue of who will do this, and on what basis.

Summary points

- There are many paths to sustainability, with a range of mixes of technology.

- Technology can serve a range of different patterns and scales of social organisation, although there may be environmental limits on what is viable.

- Choosing the best options for any particular society ought to involve a process of social negotiation, with environmental interests playing a key role.

- Environmental issues are becoming increasingly important, but it is often hard to know what are the key issues. There are many uncertainties and conflicts facing mankind should it choose to try to move towards a sustainable future.

Further reading

For an inspiring attempt to argue the case for small-scale local developments in all parts of the economy, see Richard Douthwaite's seminal *Short Circuit* (Resurgence Books/Green Books, Dartington, 1996). An updated version is available on the FEASTA web-site: http://feasta.org. On the energy technology side, the 'micropower' idea is explored in Seth Dunn's 'Micropower: the next electrical era' (Worldwatch Report 151, 2000), while a useful outline of the prospects for decentralised sustainable energy system and micro generation in industrialised countries can be found in Chris Hewitt's report 'Power to the people' produced for the Institute for Public Policy Research (IPPR) in London in 2001.

A report by the same name, but focusing on small-scale energy options for developing countries, was produced by the UK-based Intermediate Technology Development Group for the Earth Summit in 2002: http://www.itdg.org.

Certainly, sustainable energy technologies are now seen as a key to sustainable development and a sense of urgency has begun to be apparent in the publications produced by international agencies, with the emphasis on practical programmes and 'good practice'; for example, see *Sustainable Development Strategies: a*

Resource Book (Organisation for Economic Co-operation and Development (OECD) and United Nations Development Programme (UNDP), 2002).

This is a fast-moving area of technical development and social practice. To keep up to date you could make use of some of the various web-sites mentioned in the contacts list in Appendix II. The UK Department of Trade and Industry has also produced a useful 'Guide to UK renewable energy centres in the UK' (DTI/Energy 21, 2002).

 **Conclusions: the way ahead?**

- Resolving conflicts by negotiation
- Bottom-up initiatives
- Criteria for sustainability
- Human responsibilities

The way ahead for any attempt to attain sustainability must involve action at all levels, locally and globally: 'thinking globally and acting locally'. Drawing on the model of interests introduced at the start of the book, this chapter uses some of the material from the case studies and examples in the rest of the book, in order to assess some of the opportunities that exist for negotiating trade-offs between conflicting human interest and between local and global concerns. If sustainability is to be achieved, this sort of negotiation process seems vital, the final issue being whether humans are actually capable of negotiating ways to solve the environmental problems that they have created. If not, the future could be bleak.

Negotiating the future

It is not the aim of this book to provide a blueprint for what specifically should or could be done to try to solve the environmental problems that face the world. Some specific technological options have been reviewed, and it may be felt that some of them might add up to a viable sustainable energy package. However, the main point to emerge is that there is a need to choose among them and negotiate exactly how they are to be used, as part of a wider process of moving towards a sustainable future. With this process of negotiation in mind, some of the general themes and issues from earlier chapters can be usefully brought together so as to indicate some possible starting points and some possible ways in which conflicts of interest might be overcome.

The model of conflicting interests introduced in Chapter 1 highlighted the need for a process of negotiating between the various human interest groups (producers, consumers, shareholders, and the economically and politically excluded) and the interests of the planet, with the latter being given priority. Thus in Chapter 3, where we developed a list of criteria for sustainable technology, it was suggested that human sensitivities in relation to the environmental impacts of energy projects were only part of the issue. If human beings need power, then it should be human beings that paid the price. It follows from an 'eco-centric' view that the environmental costs that mattered most should not necessarily be those relating to human perceptions of the natural environment (e.g. visual intrusion) or human convenience (easy mobility in cars) as those that might affect the rest of the ecosystem.

However, there are degrees of impact: what must really be avoided are those that produce irreversible adverse changes. But in many cases it is hard to decide what the long-term effect might be of what would seem to be small-scale local impacts, especially when combined, possibly in unexpected ways, with other small impacts. Concepts like biodiversity may help to keep some sort of balance, but in the end the precautionary principle seems important: it suggests that we should avoid making any potentially dangerous changes if the likely outcomes are unknown.

While overall priority can perhaps be given to the environment, how can priorities among the various human groups be arranged? Energy projects are likely to provide benefits to some but costs to others. In Chapter 13, we looked at the issue of wind farms. The environmental benefits are clear and they accrue, in general, to everyone in terms of reduced air pollution and the avoidance of global environmental problems. But the economic benefits accrue to developers and their investors, while the environmental/amenity costs fall primarily on local residents. We suggested that one way to resolve this imbalance was to encourage local ownership of wind projects.

Unfortunately, there are not always potential solutions like this available; in some cases all the human groups may have to accept real costs in order to defend the planet. In Chapter 14 we discussed whether it would be possible to continue with economic growth indefinitely and came to the conclusion that, although it might be possible technically for the foreseeable future, there would probably also be a need for social changes, with, at the very least, less emphasis on the quantity of material consumption and more on its quality. However, by way of compensation, we noted that a switch to a more sustainable approach might help resolve the problem of unemployment – at least in some sectors of the economy.

Of course, it would not always be easy to make these changes. In Chapter 3 we looked briefly at cars and pointed out that, while in the short term consumers and producers had a common interest in continuing with existing types of vehicles, in the longer term this would not benefit either group. Everyone has to breathe and the adoption of a more sustainable approach seemed inevitable at some point.

What about the other main human interest group identified in our model – the shareholders – that is, those who benefit from investing in production using natural resources? Will profits and dividends be affected? The simplest reply is that existing patterns of commerce are environmentally unsustainable, so there is no choice, in the long term, but to make changes. Investors will have to accept their share of the burden. More subtly, it is sometimes argued that a sustainable economy could be, as profitable, or nearly, as the existing one. But it would be different – with more emphasis on what has been called 'ethical investment'.

Currently, there are many investment agencies offering shares in projects that have been subject to assessment on environmental grounds. For example, in the UK the Wind Fund is offering shares in selected wind farm projects, with a minimum stake of £300. Backed by a Danish bank, the aim is to widen the base of shareholding and to stimulate wind farm development. The local community-based wind co-ops in the Netherlands and Denmark have similar intentions.

In part these trends reflect a dissatisfaction with the traditional form of investment – whether public or private. Nationalised, state-owned companies are seen as hopelessly bureaucratic, centralised and inflexible, while the large private corporate monopolies are seen as unaccountable, insensitive and greedy. There is much that can be done to remedy this situation by way of institution innovation, increased accountability and so on at the national governmental and corporate level, but equally there is a role for local-level initiatives.

Global solutions, local initiatives

While difficult trade-offs and negotiation between human and environmental concerns will be required at all levels, it seems clear that painful negotiations can perhaps be more effective, and possibly more equitable, if carried out at the local level. That, in part, is what has been enshrined in the Local Agenda 21 programme initiated following the UN Earth Summit in 1992 (United Nations 1992).

Around the world many national and local environmental organisations have become active in this programme, including city councils and local government bodies. There are many independent local initiatives, projects and campaigns, many of which are developing practical sustainable development projects at the grass roots level: some of the organisations mentioned in the contact list at the end of this book can provide up-to-date details of what is inevitably a rapidly changing scene. But there are already many projects in the sustainable energy field around the world – East and West, North and South. For example, it is heartening to see, in the coverage provided by INforSE's journal *Sustainable Energy News*, grass roots renewable energy projects in Eastern and Central Europe.

It is also good to see local projects springing up in economically depressed areas in the West. For example, in the UK, there are projects underway designed to revitalise areas that were once reliant on coal mining, with the use of local renewable energy sources playing a key role. Earth Balance in Northumberland is making use of wind and biomass, and is hoping to create employment opportunities in an area once dominated by coal mining, while Broughton Energy Village, on an old coal mine site in the Nottingham area, is making use of wood chips from the nearby Sherwood Forest. The Earth Centre is developing an old coal mine site near Doncaster in South Yorkshire, as a demonstration of 'sustainable living' with plans to make use of local biomass as a fuel. It has benefited from support from the UK's Millennium fund.

As the director of the Earth Centre project put it, 'If sustainable development is to work, it has to start with areas like this which suffer from high unemployment (10,500 jobs have been lost in the last 10 years) and which have experienced the worst excesses of many traditional, polluting industries' (*RENEW* 1996).

As can be seen from these examples, the motto 'thinking globally and acting locally' has been taken seriously. There is much that can be done at the local level, particularly if a move can be made beyond grant-aided 'educational' projects, eco-exbibition centres and the like, to projects that meet local needs directly. Visitor centres can certainly inject tourist money into the local economy, but it would surely be more sustainable to find ways to create wealth by producing energy and other resources for local use. Some grass roots community-based initiatives may be the starting point for new enterprises, and some government support programmes can actually link to both community and entrepreneurial interests. For example, in 2002 the UK government launched a £1.6 million 'Community Renewables' initiative 'to help schools, offices and housing developments play a part in reducing the effects of climate change'. But the aim was to 'not only create environment friendly developments but to enable community groups to directly benefit from the income generated' with local employment generation via the creation of new enterprises being one possibility.

Of course there are limits to what can be done locally. As was pointed out in the previous chapter, some projects inevitably have to be larger scale and require large-scale funding of the sort only available to governments, and technical expertise only assembled in larger organisations. We are not talking about a totally decentralised system. But there are benefits from decentralisation, not least the opportunity that this would provide for more grass roots involvement with projects, including direct entrepreneurial involvement. That can ensure a degree of direct bottom-up social control, replacing the traditional top-down technocratic approach. Clearly, to make this feasible there would have to be changes in the way small enterprises, local co-ops and so on are supported, as well as changes in planning systems and in the system of local democratic control. Certainly, there will be a need to move beyond just 'consultation' on policies which have, in effect, already been decided (Elliott 2001).

Some elements of the 'bottom-up' localised approach have actually already been introduced in relation to regional planning for energy in the UK. As noted in Chapter 13, regional energy plans are being produced based on targets for the amount of renewable energy that can be produced. The overall aim is to try to match the national target, of a 10 per cent contribution from renewables by 2010, but the Regional Development Agencies are being allowed autonomy to set their own targets, reflecting local conditions. So far, the whole process has been advisory rather than mandatory, but the local planning agencies will be under pressure to come up with realistic targets and then try to meet them by adopting appropriate planning policies, for example, in relation to giving planning approval to wind farms and biomass projects.

It is not a major step from this to actually requiring agreed targets to be met with, say, access to central government funding for project development being made conditional on compliance. So they would be a combination of 'bottom-up' and 'top-down' approaches, with local autonomy and subsiduarity, combining with central policing of agreed targets.

This approach can be used to support sustainable patterns of regional economic development. For example, it is interesting to see that in Spain acceptance of wind farm plans has been made conditional on the wind turbines being manufactured within the region (Connor 2002). This has led to the growth of a distributed, rather than centralised, wind power industry in Spain. Wind turbine manufacturing is well suited to this approach – wind turbines can be manufactured efficiently in medium-sized companies.

More generally, renewables are also well suited to supporting a degree of regional autonomy. In many countries, each region could probably meet a large part of its energy needs from its own energy resources, possibly ultimately backed up by local hydrogen storage. But some areas would have much better renewable resources, and would be selling surplus energy to less well-endowed regions. So there would also be trade between regions, via the national power grid, although, as suggested above, the grid would also be topped up by power from some larger scale centralised plants.

Localisation versus globalisation

Can this sort of model be expanded further, to cover other commodities than energy? This is not the place for detailed political prescriptions for a decentralised society, but a key issue for the future must inevitably be to find a way to balance the increasing drive to globalised markets, driven by ever-increasing competitive pressures, and the need for local economic security and global environmental protection.

While, as we have seen, some people hope that it will be sufficient to impose tight controls over international trade in relation to environmental regulation, over the last few years there has been no shortage of analysis of the iniquities of globalisation and the need for an alternative approach to global – and local – economic development. Much of it stresses the need for 'localisation' – a return to an economy in which the locus of power and wealth creation is under direct local control, based on smaller scale, more self-reliant, communities, rather than on international trade overseen by giant global corporations (Hines 2000).

Part of the aim of localisation is thus to open up the economic system to the 'outsider' group in our model, that is, those who are currently excluded, dispossessed or exploited, via a new form of trade, with the emphasis being on local trade. In a fully 'localised' world economy, imports could be strictly limited to specialist goods and, similarly, exports would also be limited just to any excess volume produced – local needs, served by local markets, would be met first. That would be one way to keep predatory external companies from gaining a foothold. Indeed, it might even be that import and export taxes are imposed, with the revenue raised used to fund a national agency, linked possibly with a global agency, whose job it is to resist the growth of global monopolies – a Sustainable Trade Organization (STO) instead of a World Trade Organization. It could provide funding to make local companies effective and carefully control any cross-boundary trade. There would be an emphasis on locally owned enterprises, possibly run as co-ops, with capital raised from local sources, aided when necessary by the STO, thus giving them an edge over external, predatory companies (Dougan 2002).

This may seem a very draconian prescription – an end to 'free trade'. There will be the objection that it would fail due to the removal of the benefits of open competition, which, allegedly, ensure that prices are kept low and more efficient technologies are devised. But a well-regulated and localised sustainable economy could have economic criteria that are just as strict, but different – more concerned with resource conservation and environmental protection. Optimal environmental productivity would be the watchword, since that delivered a sustainable lifestyle for the inhabitants of the region. That would still provide an incentive for a 'Factor 4'-type technical innovation – finding ways to get more from less, with less impact.

Is this completely utopian? Maybe not. Already we are seeing attempts to introduce resource and environmental productivity considerations into economic analysis. Moreover, the 'local stewardship' scenario, which was developed for the UK government's 'foresight' exercise to contrast with its more conventional 'world markets' scenario, has elements in common with the decentralised eco-conscious future hinted at above (Foresight 2001). But could we go further and reduce or even eliminate the relentless drive to increase economic competition with other economic units such as companies and nations? That might make it possible to avoid pressure for ever-increasing exploitation of resources – a move nearer to a steady state economy. However, a weak point in this approach is that some consumers may still want the benefits of this ever-increasing exploitation of resources, regardless of its environmental impacts, and beyond the level attainable just by clever Factor 4 (doubling wealth/halving resource use) innovations. Education and normative pressures might limit rising expectations to some extent, but the only way to avoid pressures like this is for the system to deliver a lifestyle that is experienced as being qualitatively better: more satisfying, more culturally and socially enriching, more equitable and, of course, more sustainable.

Shaping the agenda

Apart from utopian visions and radical social transformations, there may still be ways in which 'thinking globally and acting locally' can ensure that local inputs and radical criteria concerning environmental decisions can be fed into the wider global debates on sustainable development. Although some of the negotiations concerning global-level concerns can only happen at international level, via representatives, even at this general policy level local input can be made by attempting to shape agendas, principles and criteria. For example, over the past few years a series of basic environmental criteria have been thrashed out, via a consensus process at the grass roots level in Sweden, under the name of the Natural Step (Greyson 1995).

The 'system criteria' that have emerged (see Box 17.1) emphasise the need for a sustainable approach in all sectors, and they have been adopted by several major Swedish companies. As can be seen there are some links with the criteria developed earlier in this book, although the Natural Step criteria are more fundamental. The Natural Step concept is now being internationalised – by inviting other organisations around the world to promote and to try to win acceptance for these basic principles. Clearly, it is a very general set of criteria, but it begins to establish a new agenda in a bottom-up way.

Natural Step is only one of several attempts to develop and promote basic criteria for sustainability. Some of the others are less general and are aimed at specific target groups, such as engineers, designers and other professionals involved with technological development. For example, in *Greening Business*, John Davis outlines a set of criteria for engineers (Davis 1991): see Box 17.2.

Box 17.1

Natural Step: system criteria

1 The use of substances from the earth's crust must not systematically increase. (This requires a considerable reduction in our dependence on mining and the use of fossil fuels.)

2 The use of substances produced by society must not systematically increase. (This requires the phasing out of persistent unnatural substances such as CFCs and PCBs. It also requires reducing the production of naturally occurring substances such as carbon dioxide and sulphur dioxide.)

3 The productivity and diversity of nature must not be systematically diminished. (This requires sweeping changes in our use of productive land, for example, in agriculture, forestry, fishing and the planning of societies.)

4 Humanity must achieve the just and efficient use of resources in society. (This requires basic human needs to be met with the most resource-efficient methods possible, including a just income distribution. Do more with less.)

Natural Step International contact point: http://www.detnaturligasteget.se/

Box 17.2

The challenge of sustainable development

Outline code of engineering for sustainable development

- The primary purpose of civil application of engineering is to harness the forces and resources of nature for the benefit of mankind and the environment.
- Benefits should be as widely available as possible, so:

  1 Natural resources should be used as efficiently as possible and renewables are preferred.
  2 The financial cost of users, over the whole life of the product, should be as low as possible.
  3 The product should not demand exceptional user skill.
  4 Production and use should not dehumanise people.

continued

- The highest technically achievable standards and energy efficiency should be aimed at.
- In evaluating the benefits/costs, long-term mass application effects must be considered as well as short-term, limited application effects.
- Every undertaking must respect human rights and human dignity. Engineering should not be carried out with the intention of advantaging some people at the expense of others; it should as far as possible advantage the disadvantaged.
- A 'total systems approach' should be adopted: maintenance, repair and reconditioning should be facilitated, and materials should be recyclable.
- Measures should be taken to prevent misuse wherever possible.
- Ultimate disposal of a product must be considered at the design stage, and plans for acceptable solutions prepared.
- Knowledge regarding safety to people and the natural environment should be freely shared.
- No relevant information regarding use/application should be withheld.
- Respect must be paid to all patented inventions and registered designs.
- Involvement in the design, development or production of illegal goods is forbidden.

Source: John Davis, 1991, *Greening Business – Managing Sustainable Development*, Basil Blackwell, Oxford

---

The advent of the concept of 'green product design' has led to a similar set of criteria emerging for designers (Blair 1992), while Mike Cooley, an exponent of the idea of 'socially useful production' has produced a set of criteria that emphasises the need to consider not only products but also the production process and the role of producers within it (Cooley 1987): see Box 17.3.

Criteria like this can play a useful role in 'raising consciousness' amongst practitioners. There is a range of other tools and processes available to help grass roots organisations to strengthen their ability to assess and influence technical and environmental decisions. A workplace example in the UK is provided by the Environmental Practitioner Programme. This consists of 100+ hours of free online learning activities that are designed to help to promote good practice in relation to environmental protection and sustainable development in the context of the ideals of Agenda 21 (see http://www.epaw.co.uk/).

At the same time there is a requirement for all the usual forms of advocacy, campaigning and lobbying to help to win wider social and

Box 17.3

Socially useful production

A tentative list of those attributes, characteristics and criteria that constitute socially useful production. It is not suggested that all these will be present in any particular socially useful product or production programme, but rather that some of these are key elements within it.

1 The process by which the product is identified and designed is itself an important part of the total process.
2 The means by which it is produced, used and repaired should be non-alienating.
3 The nature of the product should be such as to render it as visible and understandable as is possible and compatible with its performance requirements.
4 The product should be designed in such a way as to make it repairable.
5 The process of manufacture, use and repair should be such as to conserve energy and materials.
6 The manufacturing process, the manner in which the product is used and the form of its repair and final disposal should be ecologically desirable and sustainable.
7 Products should be considered for their long-term characteristics rather than short-term ones.
8 The nature of the products and their means of production should be such as to help and liberate human beings rather than constrain, control and physically or mentally damage them.
9 The production should assist co-operation between people as producers and consumers, and between nation states, rather than induce primitive competition.
10 Simple, safe, robust design should be regarded as a virtue rather than complex 'brittle' systems.
11 The product and processes should be such that they can be controlled by human beings rather than the reverse.
12 The product and processes should be regarded as important more in respect of their use value than their exchange value.
13 The products should be such as to assist minorities, disadvantaged groups and those materially and otherwise deprived.
14 Products for the Third World which provide for mutually non-exploitative relationships with the developed countries are to be advocated.
15 Products and process should be regarded as part of culture, and as such meet the cultural, historical and other requirements of those who will build and use them.
16 In the manufacture of products, and in their use and repair, one should be concerned not merely with production, but with the reproduction of knowledge and competence.

Source: Mike Cooley, 1987, *Architect or Bee?*, Chatto & Windus, London

political acceptance of the new paradigm of sustainable technological and social development that is emerging around the world.

Views as to exactly what represents the best way for individuals and organisations inevitably differ. While some people focus more on national- and international-level issues, via pressure groups and other forms of lobbying, for others local self-help initiatives and community-based campaigns are seen as the best way forward; acting locally while thinking globally. The individual's level of commitment can vary from just improving their personal lifestyles, or working with others to set up local community-based energy projects, through to changing the way that their employers or other agencies operate and campaigning on international environmental issues. (See Appendix II for a list of key contact points for futher personal involvement.)

In the final analysis, most people in the 'green' movement believe that sustainability can only be achieved if widespread agreement can be reached on what needs to be done: quite apart from being inequitable and undemocratic, draconian solutions imposed from above by elites are likely to create more problems than they solve. Grass roots initiatives may at times look weak by comparison with the scale of the problems, and the power of the large organisations that currently shape world affairs. But, as the contacts list (Appendix II) indicates, networks are growing up around the world, often making use of internet computer links, which may offer at least some solutions to the problem of thinking globally and acting locally.

New ideas?

In a presentation to a conference on alternative technology in 1994, Peter Harper, one the early pioneers of the concept, concluded as follows:

> I don't think the problem of sustainability is best solved by better technology. It probably can be – and probably will be. But this is a tiresome, bloodless, round-the-houses route when really positive and culturally deft solutions are right in front of us.
>
> (Harper 1994: 18)

He was referring to radical changes in lifestyle. He was well aware that this might threaten some people and felt that, before becoming more widely acceptable, the new types of living would probably have to be pioneered on the fringes of society, for example, in the various alternative communities that have grown up around the world. He saw the experimental residential communities like the Centre for Alternative Technology in Wales, Findhorn in Scotland and The Farm in Tennessee as 'nurseries' for developing new ways of living. At the same time, in some cases, these centres could also provide demonstrations of how alternative technologies could support the new sustainable lifestyles. The Centre for Alternative Technology, for example, is a major demonstration and exhibition site attracting over 100,000 visitors each year, but it generates most of its power from local sources – chiefly water, wind and solar power.

Not everyone will be prepared to go to such extremes. For most of us, life involves relatively conventional careers and communities. But even here changes are occurring; contemporary economic pressures seem to be forcing changes in family and community living patterns.

Perhaps most important are changes in ideas. For sustainability to mean more than just a desperate series of technical fixes, mankind seems likely to have to take on board a new view of the global ecosystem as an entity of which human beings are a part. In effect we need a new 'paradigm' – a new framework for thinking about ourselves and our relationships to the rest of the ecosystem. As Capra has suggested, we may have reached a key turning point in human development (Capra 1982).

The 1990s saw a whole range of new ideas and world visions emerge, many of them sharing an underlying holistic ecosystem view. However, there is often a lack of coherence, a degree of naivety and a potential for conflicts. Certainly, an eclectic mixture of ideas and motivations has emerged. A symbolic rejection of materialism by the affluent jostles with economic protectionism. Romantic utopianism is sometimes coupled with thinly disguised elitism. Libertarian sentiments compete with a belief in the need to reimpose 'natural order'.

Even so, it seems clear that, with the demise of some of the older ideologies that shaped the twentieth century, a new awareness is beginning to dawn, if only, as yet, on the fringes. Ecologists are developing new ideas about biodiversity and bioregionalism and our scientific understanding of ecosystems is gradually improving. In parallel, and more practically, grass roots initiatives are throwing up a new awareness of how local communities can respond to local environmental problems and new ideas are emerging from environmentalists involved with campaigns to protect endangered species and habitats. At the same time, a new appreciation of the knowledge and wisdom of native people is emerging, along with new philosophies concerning the relationship between mind, body and spirit.

The challenge for humanity

Perhaps the central issue for the future is whether a viable and widely acceptable new awareness can emerge from this complex social, cultural and in the end political process of ideological renewal, and if so, whether mankind can act on it effectively in time to avoid disaster.

Some people fear that human beings are unlikely to change their ideas in time, in which case the prospects for the future may be grim. However, some technological optimists believe that technology may enable changes to occur quickly. Some even believe that it might be possible to genetically re-engineer people to improve human mental capacities: a technical fix approach of cosmic proportions.

At the other extreme, some people ask whether there is a need to change so as to be able to take responsibility for planetary survival. The middle ground position is that mankind can and should take on a stewardship role in relation to managing the global ecosystem, since human beings have disturbed its natural functions to such an extent that it cannot sustain itself without our conscious intervention. However, some New Age thinkers see this as hopelessly arrogant and believe that mankind should abandon all pretence of being able to 'control' the situation it finds itself in, and leave everything to be resolved by the self-regulating global ecosystem, in the form of Gaia. On this view 'nature knows best', although whether the end result will be beneficial or not for human beings is less clear: as the major irritant in the system, it might be that human activities will be constrained.

Machines to the rescue?

Some New Age thinkers even seem to believe that, as Kevin Kelly has suggested in his influential book *Out of Control: the New Biology of Machines*, the future will be resolved not just by natural systems alone, but by a process of co-evolution involving the added human innovation of integrated cybernetic 'intelligent computer'-based networks. Kelly is adamant that this is not a proposal for letting machines 'take over'. For him, the world of the 'born' (i.e. living beings) and the world of the 'made' (i.e. intelligent computers and machines) will co-exist and create a new synthesis that will be able to ensure global ecosystem survival. But neither would be in control (Kelly 1995).

Kelly's idea in effect brings us full circle: if the main problem in the relationship of human beings with the rest of the ecosystem is the way our technology is used, then, for him, the answer is to let technology operate in different ways. Human beings should not use it to try to control nature, as they have tried to do in the past, but should let it help resolve the problems that they have created.

In essence, Kelly sees machines like intelligent computer systems, freed of human control, as providing the missing element in the contemporary ecosystem mix, since, unlike human beings, they can operate more like the other elements of the ecosystem, that is unconsciously and, in effect, blindly. For Kelly, the automatic, instinctive and relatively simple behaviour by the individual components or elements of complex systems, whether bee swarms, ant colonies or complete ecosystems, is the key to the survival of the system. In effect, what Kelly is arguing is that computers and integrated networks of intelligent machines can, if freed from human control, create a new layer of autonomous activity in the living self-regulating Gaian ecosystem. This has as its aim the survival of the overall ecosystem system, not the individual components, human or otherwise: the addition of the new manmade elements will simply improve the capacity of the overall system to respond to change.

In this sort of future, if it ever came about, partisan human aspirations would certainly be less of a dominant influence. But so too might the human capacity for imagination, vision, growth, co-operation and creativity, not to mention humour. In his *Architect or Bee*, Mike Cooley (1987) argues that machines can never substitute for human tacit knowledge and vision, and he quotes Karl Marx as follows: 'What distinguishes the worst architect from the best of bees is namely this. The architect will construct in his imagination that which he will ultimately erect in reality' (Marx 1974: 174).

Not surprisingly, Kelly's ideas, with his emphasis on the merits of beehive or ant colony mentality, have met with strong resistance from a range of critics, including those from the political left (Barbrook 1995). This is hardly surprising given that it could be argued that what Kelly describes already exists to some extent in the form of the global capitalist economic and industrial system. Arguably, this is unconcerned about the fate of individuals or even individual countries: its only goal is the survival, self-replication and growth of capital. Whether the vast integrated network of technology and economies that we have established can be seen in any sense as compatible with the Gaian self-regulatory ecosystem is unclear: the global techno-economic system's concern for its survival might lead it to constrain ecologically dangerous activities, but equally it could just continue blindly to the destruction of the ecosystem.

Certainly, the latter seems a strong possibility, the implication being that human beings, who created this system, must therefore try to constrain, redirect or even dismantle it. On this view, far from being a resolution of the problems created by the human use of technology, Kelly's cybernetic prescription would seem to involve forcing technology to play an even more central role, with human beings finally abandoning any pretence of responsibility.

The alternative seems clear: if cybernetic visions of automatic self-regulation do not appeal, and the global techno-economic system that mankind has created cannot be left to go its own way, then mankind will have to learn to deal with the situation itself, consciously and co-operatively.

Technical fixes can play their part. Computers can help people model, analyse and manage complex systems and computer intelligence may become a useful tool: indeed without computing power it would probably be impossible to cope with the vastness of the environmental problems that now exist (Young 1993). Biotechnology may also provide some help with some ecological problems, as may nanotechnology. For example, the technological optimists claim that genetic manipulation of crops can help develop protection against climate change, and that it may be possible to develop micro-organisms that sequester carbon dioxide or generate hydrogen (IBEA 2002). Some enthusiasts also claim that molecular-scale nano-robots can be developed to repair environmental damage, from global warming to the ozone layer (Drexler 2002). However, at the very least, the technical fixes offered by these new technologies will no doubt have the same sort of shortcomings as those we have discussed in this book in relation to earlier technical fixes. They may also open up new horrors. Many environmentalists are already worried about the impact of releasing new self-replicating transgenic biological entities into the environment. It may still be science fiction, but the impact of releasing billions of self-replicating nanobots could be even worse – well beyond human control (Miller 2002).

Few people would want to go back to a pre-technological world. However, mankind must find ways to use technology to help us avoid or reduce environmental problems. Technology must remain a tool, under human control – helping us to live on this planet without destroying it. Unless, that is, one accepts the ultimate technical fix view and believes that at some point mankind will have the technical means, as a species, to abandon this planet entirely, and start all over again somewhere else. The colonisation of space may be a worthy longer term project, as some environmentalists have argued (Deudney 1982), but for the present human beings have to try to resolve the problems they have created here, and not run away from them.

For the moment at least, there really is 'only one earth', and mankind must learn how to live on it without destroying the planet or itself.

Summary points

- Balancing the interest of people and the planet requires intervention at all levels – from the global to the local.

- Bottom-up initiatives may be able to provide some unique contributions and provide a context for experimentation with new lifestyles.

- New viewpoints on the relationship between people and the planet and new criteria for the choice of technology need to be developed.

- In the end it is the responsibility of human beings to solve the problems they have created.

Further reading

There are many ways in which you can keep abreast of the sort of developments discussed above. For example, the US-based Worldwatch Institute's annual 'State of the World' reports are a good source of data and analysis on current overall developments in the environmental policy field; it also produces reports on specific topic areas. At the grass roots level there are a whole host of journals and newsletters on green issues and local initiatives; take a look around your local alternative bookstore. The contacts list in Appendix II provides some other starting points, including the various internet-based information services.

Somewhat less radical, but usually worth a look, are the various reports produced by the UK government's Foresight programme, which seeks to look to the future with sector panels covering each area of the economy. For example, 'Fuelling the future' (2000) looked at energy options forty years ahead using four scenarios based on widely differing socio-economic futures, and this approach was adopted in the subsequent Foresight report 'Energy for tomorrow' (2001). See http://www.foresight.gov.uk.

 # Appendix I

Exploratory questions

To help you get to grips with some of the issues discussed in this book, here are some general questions that you might like to explore. Some pointers are included after each, but these are open-ended questions, designed to get you thinking about the wider issues in more detail. You may find that you will want to follow up some of the further readings mentioned at the end of each chapter.

1 *Is it really necessary to make major changes in the way energy is generated and used, or can minor adjustments suffice?*

There is a range of technical fixes on the energy use side, but it seems likely that the use of fossil fuels will be environmentally unsustainable and certainly in the longer term there will be a need for new energy sources, even if energy conservation is taken seriously. (Chapters 1, 2 and 4 provide some of the key arguments.)

2 *What are the technical options for sustainable energy supply technologies?*

Although not strictly sustainable in the longer term, nuclear power is seen by some as a possible candidate for non-fossil energy supply, but there remain problems with nuclear fission, including the waste storage issue, and fusion looks like a long shot, with its own problems. By contrast, the renewable energy sources look promising, although they too have problems. (Chapter 3 sets out the basic criteria, Chapters 5 and 6 look at nuclear power, Chapters 7 and 8 at renewables.)

3 *Can new sustainable energy technologies replace the existing range of energy technologies?*

Technically, it seems credible for renewable energy, coupled with conservation, to meet human needs into the far future, if these options are developed quickly. However, there are powerful vested interests in the technological status quo, and fossil fuel prices are relatively low, making change difficult. There is also a range of other implementation problems and a need to win public acceptance. (Chapters 9–13 review the prospects for sustainable energy and look at some of the problems.)

4 *Sustainability will never be attained unless and until major social, economic and political changes have occurred in terms of redistribution of power and wealth. Do you agree with this statement. If not what are the counter arguments?*

This is a version of the radical view that political and economic power determines all else and that tinkering on the margins, for example, with new technologies, will be of little use, since the current socio-economic system is fundamentally flawed.

A possible counter argument is that technology shapes society to some extent and if less damaging technologies can be fed in to the system, it will change. Another argument is that the system will reform itself since otherwise it will be doomed. Whether this process would benefit all concerned is, however, unclear. (Some of these issues are discussed in Chapters 14–16.)

5 *How can the necessary changes be brought about?*

Assuming that you do not feel that the situation is hopeless or that only drastic action will suffice (see question 4), the possible responses range from 'letting the market identify viable new options' (if you subscribe to the free market viewpoint) through to 'grass roots campaigning' to change viewpoints and introduce new practices and priorities. In between is a whole range of options, including personal and professional involvement with the process of developing and deploying sustainable approaches. (Chapter 17 summarises some of the options.)

6 *Is there hope for an environmentally sound future? Can sustainability be achieved?*

Mankind is technically ingenious and has also been able to develop a range of different patterns of social organisation when faced with changed circumstances. The environmental problems that lie ahead look quite serious to many people. If they are right, it remains to be seen if the necessary technical and social changes can be made sufficiently quickly to avert environmental disaster.

In terms of technology, fossil fuels remain relatively cheap and they will be with us for many decades whatever strategy is adopted. So there is much to do to ensure they are used more efficiently and cleanly. But in the years ahead pressure to phase them out is likely to grow as environmental problems mount and there will also be much to do in terms of developing the sustainable alternatives. (The contacts list in Appendix II provides some starting points if you wish to get involved.)

 # Appendix II

Contacts

There are many organisations around the world trying to support moves towards a sustainable energy future: many international environmental pressure groups like **Greenpeace** and **Friends of the Earth** have campaigns to that end, and the various national and international governmental agencies can provide information on specific policies and programmes. The UK's Department of Trade and Industry has a Sustainable Energy programme and its web-site provides updates on developments: http://www.dti.gov.uk/energy/renewables/index.shtml. The US **Department of Energy** offers similar services: see their renewable energy and energy efficiency service on the World Wide Web: http://www.eren.doe.gov/.

For independent analysis and up-to-date reports on developments you may find it useful to contact the following:

Centre for Alternative Technology, the UK's main public demonstration centre for renewable energy and allied green technologies. It also provides information on do-it-yourself approaches and has an extensive range of publications and a journal *Clean Slate*. CAT, Machynlleth, Powys SY20 9AZ, UK. Tel: +44 (0) 1654 702400. Fax: +44 (0) 1654 7002782. Web-site: http://www.cat.org.uk.

CREST, Centre for Renewable Energy and Sustainable Technology, 777N Capitol St, NE Suite 805, Washington, DC 2002, USA. Tel: +1 202 289 5370. Fax: +1 202 289 5354. e-mail: info@crest.org. World Wide Web pages: http://www.crest.org/.

See also the US renewable energy news service, Solar Access: http://www.solaraccess.com.

INforSE, the International Network for Sustainable Energy, produces a newsletter *Sustainable Energy News*, which is available electronically at http://www.inforse.org.

NATTA, the Network for Alternative Technology and Technology Assessment, c/o EERU, Open University, Milton Keynes MK7 6AA, UK. Tel: +44 (0) 1908 654638. Fax: +44 (0) 1908 653744. e-mail: S.J.Dougan@open.ac.uk.

NATTA produces a newsletter *RENEW* on technical and strategic developments in the renewable energy technology area, available on subscription. Parts of it are also available free as 'Renew On-Line' on the World Wide Web site run by the OU Energy and Environment Research Unit: http://eeru.open.ac.uk/natta/rol. html.

Rocky Mountain Institute, a centre set up by Amory and Hunter Lovins, provides information on a range of environmentally sustainable technologies, particularly in relation to energy conservation and increased energy efficiency.

RMI is at 1739 Snowmass Creek Road, Old Snowmass, CO 81654, USA. Tel: +1 303 927 3851. Fax: +1 303 927 4178. RMI can also be contacted electronically at: http://www.rmi.org.

Worldwatch Institute is at 1776 Massachusetts Ave, NW, Washington, DC 20036, USA. It publishes a range of reports and an annual 'State of the World' review. Web-site: http://www.worldwatch.org.

WISE, the World Information Service on Energy, in conjunction with the US-based **Nuclear Information Resource Service**, provides information primarily on nuclear power via its regular communiqués and reports.

WISE Amsterdam, PO Box 59636, 1040 LC Amsterdam, The Netherlands. Tel: +31 20 6126368. Fax: +31 20 6892179. Web-site: http://www.antenna.nl/wise/index.html.

 **Glossary**

acid rain acidic rain is produced as a result of the release into the atmosphere of acidic gases such as sulphur dioxide, generated by the combustion of fossil fuel in power stations and cars.

biomass biological material, such as plants. Some can be used as fuels.

end use energy the energy actually consumed at the point of use.

global warming the possible increase in average global temperatures as a result of an enhanced 'greenhouse effect' due to the release of gases such as carbon dioxide and methane into the atmosphere: global warming is one element in the resultant process of 'climate change'.

Kondratiev cycles the 'long waves' (i.e. cyclic patterns) in global economic activity identified by Kondratiev, and interpreted by some subsequent economists as being due to regular bursts of technological innovation.

nuclear fission the process of splitting the nucleus of certain atoms (e.g. uranium) with the resultant release of heat and radiation, as in atomic bombs or nuclear reactors.

nuclear fusion the process of fusing together certain light elements (e.g. hydrogen) to yield heat and radiation, as in the H-bomb and the yet to be fully developed fusion reactor.

primary energy the energy in the basic fuels or energy sources used, e.g. the energy in the fuel fed into conventional power stations.

renewable energy energy sources, such as solar energy, the winds, waves and tides, that are naturally replenished and cannot be used up. Biomass sources can also be seen as renewable, if the rate of use is matched by the rate of growth.

strategy a plan of action based on high-level goals. Strategic considerations are longer term and more fundamental than 'tactical' considerations. The concept derives from military thinking: strategy is about the overall war aims and plan while tactics are about specific battles.

sustainable development technological, economic and industrial development patterns that are environmentally and socially sustainable.

sustainable fix a more radical technical fix (see below) which may go further towards a more comprehensive and lasting solution, e.g. renewable energy.

technical fix a technical solution to a social or environmental or technical problem that tends to deal with symptoms rather than causes and often creates further problems elsewhere or at a later date.

 # References

Allen, G.R. (1991) *The Scramble for Windfarms*, Northern Devon Group of the Council for the Protection of Rural England.

Allen, P. and Todd, B. (1995) *Off the Grid*, Centre for Alternative Technology, Machynlleth.

Arnold, L. (1992) *Windscale 1957: Anatomy of a Nuclear Accident*, Macmillan, London.

Association of District Councils (1992) Report dated February 1991, cited in 'Developing wind energy: the planning issues', Report to the National Steering Committee of the Nuclear Free Local Authorities, May 1992.

Baker, C. (1991) *Tidal Power*, Peter Peregrinus/IEEE, London.

Barac, C., Spencer, E. and Elliott, D. (1983) 'Public awareness of renewable energy: a pilot study', *International Journal of Ambient Energy*, 4(4): 199–211, July.

Barbrook, R. (1995) 'What's wrong with wired', *Casablanca*, Autumn. (See also Slouka, M. (1995) *War of the Worlds*, Abacus, London.)

Beckerman, W. (1995) *Small Is Stupid*, Duckworth, London.

Beder, S. (1994) 'The role of technology in sustainable development', *Technology and Society*, 12(4): 14–19, Winter.

Beggs, C. (2002) *Energy: Management, Supply and Conservation*, Butterworth-Heinemann, Oxford.

Birkeland, J. (2002) *Design for Sustainability*, Earthscan, London.

Blair, I. (1992) 'Green products', in Charter, M. (ed.) *Greener Marketing: a Responsible Approach to Business*, Geenleaf Books, London.

Blowers, A. and Lowry D. (1991) *The International Politics of Nuclear Waste*, Macmillan, London.

Blunden, J. and Reddish, A. (eds) (1991, 1996) *Energy, Resources and Environment*, Hodder and Stoughton/Open University Press, London.

BMRB (2001) Public opinion survey carried out for the Royal Society for the Protection of Birds, British Market Research Bureau.

BNFL (2001) Submission to the PIU Energy Review by British Nuclear Fuels Ltd, Cabinet Office, London.

Bouda, Y. (1994) 'Japan's new sunshine programme', CADDET Renewable Energy Newsletter, IEA/OECD, 3, ETSU, Harwell.

Boyle, G. (eds) (1996) *Power for a Sustainable Future: Renewable Energy*, Oxford University Press/Open University Press. (New edition forthcoming.)

Boyle, G., Everett, B. and Ramage, J. (eds) (2003) *Power for a Sustainable Future: Energy Systems and Sustainability*, Oxford University Press/Open University Press.

Braun, E. (1995) *Futile Progress*, Earthscan, London.

British Wind Energy Association (1995) *Best Practice Guidelines*, BWEA, London.

Brown, L. (2001) *Wind Power: the Missing Link in the Bush Energy Plan*, Earth Policy Alert, Earth Policy Institute, Washington DC.

Brundtland, G.H. (1987) *Our Common Future*, Commission on Environment and Development, Oxford University Press, Oxford.

Calder, N. (1997) *Manic Sun*, Pilkington Press, London.

Campbell, C.J. and Laherrere, J.H. (1998) 'The end of cheap oil', *Scientific American*, March: 80–6. (See also Campbell, C., 'The imminent peak of global oil production', *FEASTA Review* 1, Foundation for the Economics of Sustainability, Dublin (http://www.Feasta.org).)

Capra, F. (1982) *The Turning Point*, Wildwood House, London.

Carley, M. and Spapens, P. (1998) *Sharing the World: Sustainable Living and Global Equity in the 21st Century*, Earthscan, London.

Carson, R. (1965) *Silent Spring*, Penguin Books, London. First published in the USA in 1962.

CATO (2002) 'Evaluating the case for renewable energy. Is government support warranted?', Jerry Taylor and Peter VanDoren. Cato Institute Policy Analysis, 422, January 2002.

Chase, R. (1998) 'Climate Change – a Role for Business' speech at the Royal Institute of International Affairs, London, November.

Clarke, A.D. (1988) 'Windfarm location and environmental impact', NATTA Report, NATTA, Milton Keynes.

—— (1994) 'Comparing the impacts of renewables', *International Journal of Ambient Energy*, 15(22): 59–72.

—— (1995) *Environmental Impacts of Renewable Energy: a Literature Review*, Technology Policy Group report, Open University, Milton Keynes.

Clarke, A.D. and Elliott, D.A. (2002) 'An assessment of biomass as an energy source: the case of energy from waste', *Energy and Environment*, 13(1): 27–55.

Connor, P. (2001) 'Wind power in Germany: the development of the German wind energy industry', Energy and Environment Research Unit, EERU Report 76, Open University, Milton Keynes.

—— (2002) 'Wind power in Spain: the development of the Spanish wind energy industry', Energy and Environment Research Unit, EERU Report 77, Open University, Milton Keynes.

Cooley, M. (1987) *Architect or Bee?*, Chatto & Windus, London.

Council for the Protection of Rural England (1991) Evidence to the Dept of Energy's Renewable Energy Advisory Group, HMSO, London.

Countryside Council for Wales (1992) 'Wind turbine power stations', CCW Report, Bangor.

County Planning Officers Society (1992) quoted in Report to the National Steering Committee of the Nuclear Free Local Authorities, May.

Davis, J. (1991) *Greening Business*, Basil Blackwell, Oxford.

Debeir, J., Deleage, J. and Hemery, D. (1991) *In the Servitude of Power: Energy and Civilisation through the Ages*, Zed Books, London and New Jersey.

Department for Environment, Food and Rural Affairs (2001) *Government Looks for Public Consensus on Managing Radioactive Waste*, Press release, DEFRA, London, 12 September.

Department of Energy (1989) 'The Severn barrage project: general report', Energy Paper 57, HMSO, London.

Department of the Environment (1992) Secretary of State's response to Inspectors' recommendations, quoted in *Daily Telegraph*, 26 March.

Department of Trade and Industry (1992) Government's response to the Energy Select Committee's Report on Renewable Energy, DTI, 16 July.

—— (1993) *Digest of United Kingdom Energy Statistics 1992*, HMSO, London.

—— (1994a) 'Outcome of the review of statistical methodologies for the compilation of overall energy data', *Energy Trends* (HMSO, London), July.

—— (1994b) 'New and renewable energy: prospects for the UK', Energy Paper 62, DTI, HMSO, London.

—— (1995) *Digest of United Kingdom Energy Statistics*, HMSO, London.

—— (1996) *'Energy Trends' Statistical Bulletin*, HMSO, London, January.

—— (2000) *The Digest of United Kingdom Energy Statistics 2000*, HMSO, London.

—— (2002) 'Report on Public Consultation on attitudes to energy and the environment', DTI, London, October.

Department of Trade and Industry/Energy 21 (2002) 'Guide to UK Renewable Energy Centres in the UK', DTI/Energy 21, London.

DETR (2000) 'Climate change: UK Programme', Department of the Environment, Transport and the Regions, HMSO, London. (See also DETR 'Guidance on preparing regional sustainable development frameworks.)

Deudney, D. (1982) 'Space: the high frontier', Worldwatch Paper 50. (See also Kim Stanley Robinson's science fiction novel *Green Mars* (HarperCollins, London, 1994), for an interesting exploration of some of the issues.)

Devall, B. (1988) *Simple in Means, Rich in Ends: Practicing Deep Ecology*, Peregrine Smith Books, Salt Lake City, UT.

Devall, B. and Sessions, G. (1985) *'Deep Ecology'*, Peregrine Smith Books, Salt Lake City, UT.

Dickson, D. (1974) *Alternative Technology: the Politics of Technical Change*, Fontana, London.

Dobson, A. (1995) *Green Political Thought*, Routledge, London.

Donaldson, D. and Betteridge, G. (1990) 'Carbon dioxide emissions from nuclear power – a critical analysis of FOE 9', *Atom*, 400, February 18–22.

Donaldson, D., Tolland, H. and Grimston, M. (1990) *Nuclear Power and the Greenhouse Effect*, UK Atomic Energy Authority, January.

Dougan, T. (1998) 'Sustainable housing: some lessons from Milton Keynes', NATTA report, Milton Keynes. (See also Dougan, T. (1998) 'Green housing', *RENEW* 114, July–August: 18–21; Dougan, T. (1998) 'Building sustainability', *Building for a Future*, 8(2): 39–43.)

—— (2002) 'Which sustainable future?' *RENEW* 137, May–June: 23. (See also Dougan, T. (2002) 'Back to the future' *RENEW* 139, September–October: 18–22.)

Douthwaite, B. (2002) *Enabling Innovation*, Zed Books, London.

Douthwaite, R. (1996) *Short Circuit*, Resurgence Books/Green Books, Dartington.

Drexler, K.E. (2002) *Engines of Creation*, Anchor Books, Peterborough. (See Web and PDF versions (2002) adapted by R. Whitaker at http://www.foresight.org/EOC/.)

Dunn, S. (2000) 'Micropower: the next electrical era', Worldwatch Paper 151, Worldwatch Institute, Washington, DC.

Eggar, T. (1993) DTI press release, 15 November.

—— (1994) DTI press release, 11 March.

Ekins, P. (1999) *Economic Growth and Environmental Sustainability*, Routledge, London.

—— (2001) 'The UK's transition to a low carbon economy', Forum for the Future, London.

Elgin, D.S. and Mitchel, A. (1977) 'Voluntary simplicity: life style of the future?', *Futures*, 11(4): 200–6. (See also Elgin, D.S. (1981) *Voluntary Simplicity*, William Morrow, New York.)

Elliott, D. (1978) *The Politics of Nuclear Power*, Pluto Press, London.

—— (1989) 'Privatisation, nuclear power and the trade union and labour movement', Technology Policy Group Occasional Paper 19, Open University, Milton Keynes, November. (See also three earlier TPG reports on this topic, TPG Occasional Paper 3 (1981), 14 (1987) and 17 (1988).)

—— (1992) 'Renewables and the privatisation of the electricity supply industry: a case study', *Energy Policy*, 20(3): 257–68, March.

—— (1995) 'The UK wave energy programme', Technology Policy Group Occasional Paper 27, Open University, Milton Keynes, March.

—— (1996) 'Renewable energy policy in the UK: the limits of the market approach', Technology Policy Group Occasional Paper 29, Open University, Milton Keynes, March.

—— (2001) 'Risk governance: is consensus a con?', *Science as Culture*, 10(2): 265–71.

Elliott, D. and Elliott, R. (1976) *The Control of Technology*, Wykeham, London.

Elliott, D. and Toke, D. (2000) 'A fresh start for windpower?', *International Journal of Ambient Energy*, 21(2): 67–76.

Elliott, J. (1999) *An Introduction to Sustainable Development*, Routledge, London.

Energy Technology Support Unit (1982) 'Strategic review of renewable energy technologies', vol. 1, ETSU Report R13, Harwell.

—— (1993a) 'Tidal stream energy review', ETSU Report T/05/00155/REP, Harwell.

—— (1993b) 'Attitudes towards windpower: a survey of opinion in Cornwall and Devon', ETSU Report W/13/00 354/038/REP, Harwell.

—— (1994) 'An assessment of renewable energy for the UK', ETSU Report R82, Harwell.

England, G. (1978) 'Renewable sources of energy – the prospects for electricity', ATOM, 264, UKAEA, October: 270–2.

European Parliament (1999) Directorate General for Research 'Emerging nuclear energy systems, their possible safety and proliferation risks', Working Paper, Energy and Research Series, ENER 111EN.

Evans, D.L. (1991) 'Dyfi windfarm worries', Rural Wales, Campaign for the Protection of Rural Wales, Spring .

EXTERNE (2001) 'Externalities of Energy' reports on the EXTERNE programme, European Commission, DG12, L-2920 Luxembourg, 1995, 1999, 2001.

Eyre, B. (1991) 'The longer term direction for the nuclear industry', ATOM, 411, UKAEA, March: 8–14.

Eyre, N. (1993) 'Comparison of environmental impacts of electricity from different sources', Nuclear Energy, 32(5): 321–6.

Farinelli, U. (1994) 'A setting for the diffusion of renewable energies', Renewable Energy, 5(1): 77–82.

Flavin, C. (1995) Power Surge: Guide to the Coming Energy Revolution, Worldwatch Institute, Washington DC.

Flood, M. (1991) Energy Without End, Friends of the Earth, London.

Foresight (2001) 'Energy for tomorrow: powering the 21st century', Energy Futures Task Force, Foresight Programme, Department of Trade and Industry, London.

Freeman, C. (1992) The Economics of Hope, Pinter, London.

Friends of the Earth (1994) Planning for Wind, FoE, London.

Galli, R. (1992) 'Structural and institutional adjustments and the new technological cycle, Futures, October: 775–88.

Gallup (1994) Gallup poll data from the UK Government's Nuclear Review, Submission by Stop Hinkley Expansion, September.

Garrison, J. (1980) From Hiroshima to Harrisburg, SCM Press, London.

Garvin, R.L. (1998) 'The nuclear fuel cycle', Pugwash Conference, Paris, December.

Genus, A. (1993) 'The political construction and control of technology: wave power renewable energy technologies', Technology Analysis and Strategic Management, 5(2): 137–49.

Gipe, P. (1995) Windpower Comes of Age, Wiley, London.

Gottlober, P., Steinert, M., Weiss, M. et al. (2001) 'The outcome of local radiation injuries: 14 years of follow-up after the Chernobyl accident', *Radiation Research*, 155(3): 409–16. (See also Rashad, S.M. and Hammad, F.H. (2000) 'Nuclear power and the environment: comparative assessment of environmental and health impacts of electricity-generating systems', *Applied Energy*, 65(1–4): 211–29.)

Greenpeace (1992) Evidence to the House of Commons Select Committee on Energy, hearings on Renewable Energy, Session 1991–2, Fourth Report, vol. II, pp. 157–67.

—— (1993) 'Towards a fossil free energy future', Stockholm Institute report for Greenpeace International, London, April.

—— (1995) Evidence on the CEC Green Paper to the House of Lords Select Committee on the European Communities, Session 1994–5, 17th Report 'European Union Energy Policy' vol. II, Evidence, p. 103, HMSO, London, July.

—— (1997) 'Stop stoking the fire', Greenpeace International, London, based on a report for Greenpeace on 'Energy subsidies in Europe', produced by Vrije University, Amsterdam.

Greyson, J. (1995) 'Taking a natural step', *New Ground*, 45, SERA, London.

Grimston, M. and Beck, P. (2000) *Civil Nuclear Energy: Fuel for the Future or Relic of the Past*, Royal Institute of International Affairs, London.

—— (2002) *Double or Quits?: the Global Future of Civil Nuclear Energy*, Earthscan/Royal Institute of International Affairs, London.

Grob, G. (1994) 'Transition to the sustainable energy age', *European Directory of Renewable Energy Suppliers and Sources*, James and James, London.

Grubb, M. (1991) 'The integration of renewable electricity sources, *Energy Policy*, 19(7): 670–88, September.

Grubb, M., Vrolijk, C. and Brack, D. (1999) *The Kyoto Protocol: a Guide and Assessment*, Earthscan, London.

Hain, P. (2001) Peter Hain, then Energy Minister, in a House of Commons Adjournment debate on Renewable Energy, Westminster Hall, 5 April.

Hammonds, A. (1998) *Which World? Scenarios for the 21st Century*, Earthscan, London.

Hansen, J. *et al.* (2002) 'Climate forcings in Goddard Institute for Space Studies SI2000 simulations', *Journal of Geophysical Research*, DOI10.1029/2001JD001143.

Harper, M. (1994) 'Wind energy still blowing strong', *Safe Energy*, 99, February/March: 13.

Harper, P. (1974) 'What's left of AT', *Undercurrents*, 6: 35–8.

—— (1994) 'The "L" word: AT and lifestyles', Paper presented to the EERU 'AT 2000' Conference, Conference Proceedings, Energy and Environment Research Unit, Open University, Milton Keynes.

—— (1996) 'Getting through to Sweden', *Clean Slate*, 20: 10–11, Centre for Alternative Technology, Alternative Technology Association, Machynelleth.

Heap, B. and Kent, J. (eds) (2000) *Towards Sustainable Consumption: a European Perspective*, Royal Society, London.

Herring, H. (1999) 'Does energy efficiency save energy? The debate and its consequences', *Applied Energy*, 63: 209–26. (See also Herring, H. (2000) 'Is energy efficiency environmentally friendly?', *Energy and Environment*, 11(3): 313–25.)

Hewitt, C. (2001) 'Power to the people', Institute for Public Policy Research, London.

Hines, C. (2000) *Localisation: a Global Manifesto*, Earthscan, London.

Hohmeyer, O. (1992) 'Renewable energy and the full cost of energy', *Energy Policy*, 20(4): 365–75, April.

Hongpeng, L. (2000) 'Renewable energy development strategy and market potential in China', World Renewable Energy Congress VI Congress Papers, Pergamon Press, 90–96.

Horwood, T. (1992) quoted in *Windpower Monthly*, November.

Houghton, J. (1997) *Global Warming: the Complete Briefing*, Cambridge University Press, Cambridge.

Houlden, V. (2000) 'Coolants spark heated debate', *Financial Times*, 26 July.

Hoymer, O. (1988) *The Social Costs of Energy Consumption*, Springer, Berlin.

Hunt, D. (1991) Planning consent letter and Annex (P62/546), Welsh Office, Cardiff, 19 September.

Hyam, P. (1995) 'Talking energy', *New Review*, 26, Department of Trade and Industry, August.

Hydro Quebec (2000) 'Comparing environmental impacts of power generation options', Fact Sheets, in PDF from http://www.hydroquebec.com/environment/. (See also Gagnon, L., Belanger, C. and Uchiyama, Y. (2002) 'Life cycle assessment of electricity generation options: the status of research in the year 2001', *Energy Policy*, 30: 1267–78.)

IAEA (2000) International Atomic Energy Agency data as reported on the Word Nuclear Associations web-site at http://www.world-nuclear.com/education/ueg.htm.

IBEA (2002) Institute for Biological Energy Alternatives, Washington, DC, press release on $3m Dept of Energy grant, 21 November.

IEA (2001a) 'World energy outlook', International Energy Agency, Paris.

—— (2001b) 'Energy technology R&D statistics', International Energy Agency, Paris.

IPCC (1995) 'Second assessment of scientific-technical information relevant to interpreting Article 2 of the UN Framework Convention on Climate Change', Intergovernmental Panel on Climate Change, Geneva.

—— (2001) 'Third assessment: climate change 2001: synthesis report, summary for policymakers', Intergovernmental Panel on Climate Change, Geneva.

Jackson, K. (2002) Sir Ken Jackson, then General Secretary of Amicus, quoted in *Utility Week*, 21 June.

Jackson, T. (1992) 'Renewable energy: summary paper for the renewable series', *Energy Policy*, 20(9): 861–83.

Jeffrey, K. (1995) *Independent Energy Guide*, Chelsea Green, VT.

Johansson, T., Kelly, H., Reddy, A. and Williams, R. (1993) *Renewable Energy: Sources for Fuels and Electricity*, Earthscan, London.

Kahn, H. and Simon, J. (1985) *The Resourceful Earth*, Blackwell, Oxford.

Karnoe, P. (1990) 'Technological innovation and industrial organisation in the Danish wind industry', *Entrepreneurship & Regional Development*, 2: 105–23.

Keen, B.E. and Maple, J.H.C. (1994) 'JET and nuclear fusion', JET Publications Group, Culham, June. (See also the leaflet 'Energy: the importance of fusion research', JET Publications, Culham.)

Kelly, K. (1995) *Out of Control: the New Biology of Machines*, Fourth Estate, London.

Kemp, D. (1994) *Global Environmental Issues*, Routledge, London.

Key, C. (1994) quoted in the *Sunday Telegraph*, 27 February.

Kidd, S. (2000) 'Can Nuclear Power succeed in Liberalised Electricity markets', Energy Resources e-conference organised by the World Energy Council.

Krewitt, W. (2002) 'External costs of energy – do the answers match the questions? Looking back at 10 years of ExternE', *Energy Policy*, 30: 838–48.

Laughton, M. (2002) 'Renewables and UK grid infrastructure', *Power in Europe*, 383, September.

Legget, J. (1991) 'Energy and the new politics of the environment', *Energy Policy*, 19(3): 161–71, March.

Lewis, H., Gertsakis, J., Grant, T., Morelli, N. and Sweatman, A. (2001) *Design + Environment: A Global Guide to Designing Greener Goods*, Greenleaf, Sheffield.

Llwyd Evans, D. (1991) 'Dyfi windfarm worries', *Rural Wales*, Campaign for the Protection of Rural Wales, Spring .

Lomborg, B. (2001) *The Skeptical Environmentalist*, Cambridge University Press, Cambridge.

Lovelock, J. (1979) *Gaia: a New Look at Life on Earth*, Oxford University Press, Oxford. (See also Lovelock, J. (1988) *The Ages of Gaia*, Oxford University Press, Oxford.)

Lovins, A. (1977) *Soft Energy Paths: Towards a Durable Peace*, Penguin, Harmondsworth.

McDonald, A. and Schrattenholzer, L. (2001) 'Learning rates for energy technologies', *Energy Policy*, 29: 255–61.

McKenzie, D. (1991) *Green Design: Design for the Environment*, Lawrence King, London. (See also *EcoDesign*, 4(1), 1996.)

Macpherson, G. (1995) *Home Grown Energy*, Farming Press, Ipswich.

McSorley, J. (1990) *Living in the Shadow*, Pan Books, London.

Marx, K. (1974) *Capital*, Lawrence and Wishart edition, vol. 1. Original German edition 1867.

Meadows, D., Meadows, D. and Randers, J. (1972) *Limits to Growth*, Earth Island, London.

—— (1992) *Beyond the Limits: Global Collapse or a Sustainable Future?*, Earthscan Publications, London.

Medvedev, Z. (1990) *The Legacy of Chernobyl*, Basil Blackwell, Oxford. (See also 'Chernobyl ten years on', *Safe Energy* 108, March–May 1996; Tucker, A. 'Chernobyl ten years on', *Guardian*, 17 February 1996.

Mellor, D. (1982) Parliamentary Answer, 31 March.

Mensch, G. (1979) *Stalemate in Technology*, Ballinger, Cambridge, MA. (German original published in 1975.)

Meridian (1998) 'Energy system emissions and material requirements', prepared for the Deputy Assistant Secretary for Renewable Energy, US Department of Energy, Washington, DC, by the Meridian Corporation.

Meyer, A. (2000) 'Contraction and convergence – the global solution to climate change', Schumacher Society Briefing Number 5, Greenbooks, Dartington.

Milborrow, D. (2001) 'Penalties for Intermittent Sources of Energy', Working paper for the PIU Energy Review, Cabinet Office, London (http://www.piu.gov.uk/2002/energy/report/working%20papers/Milborrow.pdf).

Miller, P. (2002) 'Big, big plans for very small thing', *Green Futures* 35(2): 50–1.

Moore, J. (1980) At the opening of the Southampton wave test tank, UK Department of Energy press release, 26 September.

Morris, J. (2002) *Sustainable Development: Promoting Progress or Perpetuating Poverty*, Profile Books, London.

Mortimer, N. (1990) 'The controversial impact of nuclear power on global warming', NATTA Discussion Paper 9, NATTA: Milton Keynes, September.

—— (1991) 'Energy analysis of renewable energy sources', *Energy Policy*, 19(4): 374–85, May.

NALGO (1990) 'Greenprint for action', National and Local Government Officers Union/National Extension College course pack.

NAO (1993) 'The cost of decommissioning nuclear facilities', National Audit Office, London.

—— (1994) 'The renewable energy research, development and demonstration programme', National Audit Office, London.

NATTA (1993a) *RENEW*, 85, September–October.

—— (1993b) 'The windfarm debate', evidence to the Welsh Affairs Committee, vol. III of the Committee's report (WE52).

—— (1994a) 'Story of ugliness of wind turbines: wind turbines – the truth', unsigned 'Factsheet' dated 1993, reprinted in *RENEW* 87, NATTA, Milton Keynes, January 1994.

—— (1994b) *Windfarm Backlash*, compilation of press cuttings, vols I and II, NATTA, Milton Keynes.

—— (1994c) DTI press release on National Wind Power programme, quoted in *RENEW* 87, NATTA, Milton Keynes, January 1994.

—— (2001) 'Public attitudes towards wind farms in Scotland', System Three Social Research (as reported in *RENEW* 130, NATTA, Milton Keynes).

Nieuwenhurst, P. Cope, P. and Armstrong, J. (1992) *The Green Car Guide*, Green Print.

NII (2001) *A Review by HM Nuclear Installations Inspectorate of the British Energy Plc's Strategy for Decommissioning of its Nuclear Licensed Sites*, London, NII.

NIRS (2000) *'CDM – a New Nuclear Subsidy?' Climate Change and the CDM Briefing Note*, Washington, DC, Nuclear Information and Resource Service.

Nordhaus, W. (1997) *The Swedish Nuclear Dilemma: Energy and the Environment,* Resources for the Future, Washington, DC.

Norgard, J.S., Nielsen, P.S. and Viegand, J. (1994) 'Low electricity Europe', in A.T. de Almeida *et al.* (eds), *Integrated Electricity Resource Planning*, Kluwer Academic Publishers, Lancaster, 399–417. See also Norgard, J. and Christensen, B. (1993) 'Towards sustainable energy welfare', *Perspectives in Energy* 2: 313–32.

O'Connor, J. (1991) *Capitalism, Nature, Socialism*, (CNS) Conference Papers, CES/CNS Pamphlet 1, Centre for Ecological Socialism, Santa Cruz.

OECD and NEA (1989) 'Plutonium fuel: an assessment', Report by an Expert Group, Paris, Office of Economic Co-operation and Development and Nuclear Energy Agency.

OECD/UNDP (2002) *Sustainable Development Strategies: a Resource Book*, Organisation for Economic Co-operation and Development (OECD) and United Nations Development Programme (UNDP).

Office of Science and Technology (1995) Technology Foresight: Progress through Partnership, vol. 13, *Energy*, OST, HMSO, London.

Olivier, D. (1996) 'Energy efficiency: key to a sustainable energy future', NATTA Technical Paper 15, Milton Keynes.

Olivier, D., Willoughby, J. *et al.* (1996), 'Review of ultra-low-energy homes' General Information Reports 38 and 39. BRECSU, Building Research Establishment, Watford.

Pasqualetti, M., Gipe, P. and Righter, R. (eds) (2000) *Wind Power in View: Energy Landscapes in a Crowded World*, Academic Press/Elsevier, London.

Patterson, W. (1976) *Nuclear Power*, Penguin, Harmondsworth.

Peake, S. (1994) *Transport in Transition: Lessons from the History of Energy*, Royal Institute of International Affairs, London.

Pena-Torres, J. and Pearson, J.G. (2000) 'Carbon abatement and new investment in liberalised electricity markets: a nuclear revival in the UK', *Energy Policy*, 28: 115–35.

Pepper, D. (1984) *The Roots of Modern Environmentalism*, Croom Helm, London.

PIU (2002) 'The energy review', Performance and Innovation Unit, Cabinet Office, London.

Porteous, A. (1992) 'Energy and waste', NATTA Technical Paper 7, NATTA, Milton Keynes.

—— (2000) 'Dioxins', in *Dictionary of Environmental Science and Technology*, 3rd edn, Wiley, London.

Ramage, J. (1993, 1997) *Energy – a Guidebook*, Oxford University Press, Oxford.

Read, P.R. (1993) *Ablaze: the Story of Chernobyl*, Secker and Warburg, London.

Redlinger, R., Andersen, P. and Morthorst, P. (2002) *Wind Energy in the 21st Century Economics, Policy, Technology, and the Changing Electricity Industry*, Palgrave/UNEP, Basingstoke.

RENEW (1996) 'The earth centre', *RENEW*, 100, NATTA, Milton Keynes, March–April. (See also *EcoDesign*, 3(4), 1995.)

Renewable Energy Advisory Group (1992) Report to the DTI on Renewable Energy, Energy Paper 60, HMSO, London.

Ridley, M. (1994) *Daily Telegraph*, 6 February.

Rifken, J. (2002) *The Hydrogen Economy*, Polity, Cambridge.

Rogner, H. and Langlois, L. (2000) 'Market reform and its impacts on nuclear power', Energy Resources e-conference organised by the World Energy Council.

Rosenberg, N. and Fristak, C.R. (1983) 'Long waves and economic growth: a critical apprasial', *American Economic Review*, 73(2): 146–51.

—— (1994) 'Technological innovation and long waves', in Rosenburg, N. (ed.), *Exploring the Black Box*, Cambridge University Press, Cambridge.

Ross, D. (1995) *'Power' from the waves*, Oxford University Press, Oxford.

—— (2002) 'Scuppering the waves: how they tried to repel clean energy', *Science and Public Policy*, 29(1): 25–35, February.

Royal Commission on Environmental Pollution (1976) 'Nuclear power and the environment' (The Flowers Report), 6th Report, Cmnd 6618, HMSO, London.

—— (2000) 'Energy – the changing climate', 22nd Report, Cmnd 4749, HMSO, London.

Royal Society (2001) 'The role of land carbon sequestration in mitigating global climate Change', Royal Society report, London.

Royal Society/Royal Academy of Engineering (1999) 'Nuclear energy: the future climate', Royal Society/Royal Academy of Engineering, London.

Ryan, C. (1994) 'The practicalities of ecodesign', in Harrison, M. (ed.), *Ecodesign in the Telecommunications Industry*, RSA Environmental workshop, 3–4 March 1994, RSA, London.

Safe Alliance (1992) 'Biofuels and European agriculture', Sustainable Agriculture, Food and Environment (SAFE), London.

Salter, S. (1981a) quoted in the *Financial Times*, 18 May 1981.

—— (1981b) 'Wave energy: problems and solutions', *Journal of the Royal Society of Arts Proceedings*, London, August.

Scheer, H. (1994) *Solar Manifesto*, James and James, London.

—— (2002) *The Solar Economy: Renewable Energy for a Sustainable Global Future*, Earthscan, London.

Schmidheiny, S. (1992) *Changing Course*, Business Council for Sustainable Development, MIT Press, Cambridge, MA.

Schoon, N. (1989) *Independent*, 10 July.

Schumacher, F. (1973) *Small Is Beautiful*. (First published 1973; many editions subsequently, e.g., Hutchinson, London, 1980.)

Schumpeter, J.A. (1939) *Business Cycles*, McGraw-Hill, New York.

Select Committee (1984) House of Commons Select Committe on Energy, Session 1983–4, Ninth Report, 'Energy R, D & D in the UK', HMSO, London.

—— (1992) House of Commons Energy Committee, Session 1991–2, Fourth Report, 'Renewable energy', vol. III, March, HMSO, London.

—— (2001) House of Commons Select Committee on Science and Technology, Session 2000–1, Seventh Report, 'Wave and tidal energy', HC 291, April, HMSO, London.

—— (2002) House of Commons Select Committee on Trade and Industry, Session 2001–02, Second Report, 'Security of energy supply', HC364-I, HMSO, London.

Sessions, G. (1996) *Deep Ecology for the Twenty-First Century*, Schumacher Books, Dartington.

Shell (1995) 'The evolution of the world's energy system 1860–2060', Shell International, London.

—— (2001) 'Exploring the future: energy needs, choices and possibilities – scenarios to 2050', Shell International, London.

Smil, V. (1999) *Energies: an Illustrated Guide to the Biosphere and Civilization*, MIT Press, Cambridge, MA.

Smith, A. (2002) 'Transforming technological regimes for sustainable development: a role for appropriate technology niches?', Electronic Working Paper 86, Environment & Energy Programme, SPRU, University of Sussex, Brighton (http://www.sussex.ac.uk/spru/).

Sorenson, B. [1979] (2001) *Renewable Energy*, Academic Press, London.

Stirling, A. (1994) 'Diversity and ignorance in electricity supply investment', *Energy Policy*, 22(3): 195–216, March.

STOA (2001) 'Possible toxic effects from the nuclear reprocessing plants at Sellafield (UK) and Cap de La Hague (France)', WISE-Paris report for the European Parliament Scientific and Technological Options Assessment programme, Strasbourg.

Stoddard, W. (1986) 'The California experience', Paper to the Danish Wind Energy Association Conference, quoted in Karnoe, P. (1990) 'Technological innovation and industrial organisation in the Danish wind industry', *Entrepreneurship and Regional Development*, 2.

Thayer, R. and Hanson, H. (1988) 'Wind on the land', *Landscale Architecture*, March. (See also: Thayer, R. and Freeman, C. (1987) 'Altamont: public perceptions of a wind energy landscape', *Landscape and Urban Planning*, 14: 379–88; Pasqualetti, M.J. and Butler, E. (1987) 'Public reaction to wind development in California', *International Journal of Ambient Energy*, 8(2): 83–90, April.)

Thorpe, T. (1992) 'A review of wave energy', ETSU Report R-72, Energy Technology Support Unit, Harwell.

Tickell, C. (1994) 'Concepts and dilemmas of sustainable development', in N. Steen (ed.), *Sustainable Development and the Energy Industries*, Earthscan, London.

Toke, D. (2002) 'Wind power needs CHP', *RENEW*, 141, January–February, 25.

Trainer, T. (1995) *The Conserver Society*, Zed Press, London.

TUC (1989) 'Green charter', Trades Union Congress, London.

Twidell, J. and Wier, D. (1990) *Renewable Energy Resources*, E. & F. N. Spon, London.

Tylecote, A. (1991) *The Long Wave in the World Economy*, Routledge, London.

United Nations (1992) Agenda 21 Press Summary, United Nations Conference on Environment and Development, Department of Public Information, UN, New York. (See also UN Committee on Environment and Development, 'Connections', UNCED-UK Newsletter, UN Association, London.)

UNDP and UNICEF (2002) *The Human Consequences of the Chernobyl Nuclear Accident*, United Nations Development Programme and UN Children's Fund.

UNSCEAR (2000) Report to the General Assembly by the UN Scientific Committee on the Effects of Atomic Radiation, New York.

Victor, D.G. (2000) *The Collapse of the Kyoto Protocol and the Struggle to Slow Global Warming*, Princeton University Press, Princeton, NJ.

von Weizsacker, E., Lovins, A. and Lovins, H. (1994) *Factor Four*, Earthscan, London.

Wainwright, H. and Elliott, D. (1982) *The Lucas Plan: a New Trade Unionism in the Making*, Allison and Busby, London.

Walker, S. (1993) 'Down on the windfarm', NATTA report, Milton Keynes.

Wallace, D. (1996) *Sustainable Industrialisation*, Earthscan, London.

Watson, W. (1994) 'Integrated tidal power', NATTA Technical Paper 10, NATTA, Milton Keynes.

WEA (2000) *World Energy Assessment: Energy and the Challenge of Sustainability*, UN Development Programme, UN Department of Economic and Social Affairs and the World Energy Council.

Welford, R. and Starkey, R. (eds) (2001) *The Earthscan Reader in Business and Sustainable Development*, Earthscan, London.

Welsh Affairs Committee (1994), Session 1993–4, Second Report, 'Wind energy', vol. 1, HMSO, London, July.

Wilson, B. (2002) Speech at the launch of the DTI supported ESRC Sustainable Technologies Programme, British Academy, London, 13 November.

Wollard, R. and Aleck Ostry, A. (2000) *Fatal Consumption: Rethinking Sustainable Development*, UBC Press, Vancouver.

World Commission on Dams (2001) 'Dams and development: a new framework for decision-making', London.

World Energy Council (1994) *New and Renewable Energy Resources: a Guide to the Future*, Kogan Page, London.

—— (1995) 'Energy for our common world – what will the future ask of us?', Conclusions and Recommendations, 16th WEC Congress, Tokyo.

WWF (2002) 'The American way to Kyoto', World Wide Fund for Nature, Washington, DC.

Young, J. (1993) 'Global network: computers in a sustainable society', Worldwatch Paper 115, Worldwatch, Washington, DC.

Index

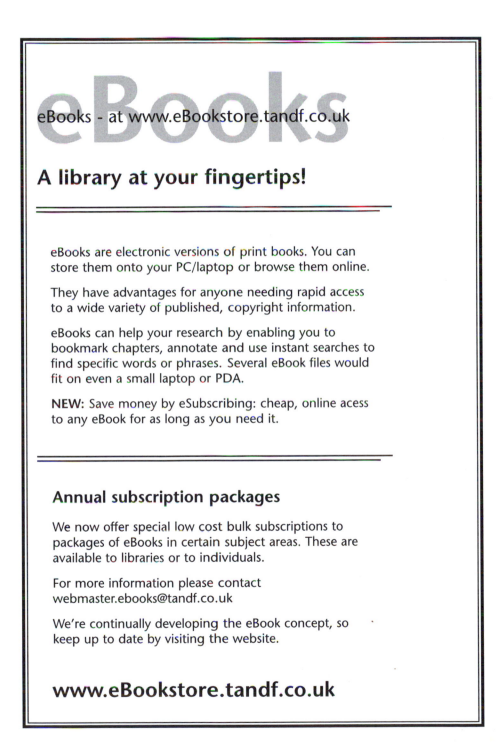